Karl Schügerl

Solvent Extraction in Biotechnology

Recovery of Primary and Secondary Metabolites

With 130 Figures and 48 Tables

Springer-Verlag

Berlin Heidelberg New York
London Paris Tokyo
Hong Kong Barcelona Budapest

Prof. Dr. Dr. h. c. Karl Schügerl

University of Hannover
Institut für Technische Chemie
Callinstraße 3
30167 Hannover / FRG

ISBN 3-540-57694-0 Springer-Verlag Berlin Heidelberg New York
ISBN 0-387-57694-0 Springer-Verlag New York Berlin Heidelberg

Library of Congress Cataloging-in-Publication Data
Schügerl, K. (Karl)
Solvent extraction in biotechnology : recovery of primary and sencondary metabolites /Karl
Schügerl.
Includes index.
ISBN 0-387-57694-0 (alk. paper: US). -- ISBN 0-387-57694-0 (Germ.)
1. Extraction (Chemistry) 2. Biotechnology--Methodology. 3. Metabolites--Separation. I. Title.
TP248.25.E88S38 1994 660'.284248--dc20

Typesetting: Macmillan India Ltd., Bangalore-25
Offsetprinting: Saladruck, Berlin; Bookbinding: Lüderitz & Bauer, Berlin
SPIN: 10058372   02/3020   5 4 3 2 1 0   Printed on acid-free paper

Preface

Two excellent books on extraction are available:

→ Handbook of solvent extraction T.C. Lo, M.H.I. Baird and C. Hanson, eds., John wiley & Sons, New York, 1983, and

→ "Science and Practice of Liquid-Liquid Extraction" by J.D. Thornton, ed., Clarendon Press Oxford, 1992, which also consider extraction in biotechnology, but in only one chapter. On the whole, however, these chapters cover only a very small part of relevant biotechnological aspects.

On the other hand, several books deal with separation in biotechnology: E.g.,

"Bioseparations. Downstream Processing for Biotechnology" by P.A. Belter, E.L. Cussler and W.S. Hu, John Wiley & Sons, New York, 1988,

"Separation and Purification Techniques in Biotechnology" by F.J. Dechow, Noyes Publications, New Yersey, 1989, and

"Separations for Biotechnology" M.S. Verral and M.J. Hudson, eds., Ellis Horwood Ltd., Chichester, 1987.

However, none of them treat extraction in detail.

Also excellent handbooks:

"Biochemical Engineering and Biotechnology Handbook" by B. Atkinson and F. Mavitune, The Nature Press, New York, 1983, and

"Biotechnology of Industrial Antibiotics" E.J. Vandamme, ed., Marcel Dekker, Inc., New York, 1984,

consider only special cases of extraction and do not go into detail.

On account of our laboratory and pilot plant experience with the recovery of primary and secondary metabolites by extraction and the invitation of the Springer Verlag to participate in publishing with an actual topic, I decided to write this book. Particular attention was paid to modern extractants with high selectivity and modeling of extraction processes.

This volume should provide the reader with a comprehensive overview of this subject and reference material for students of biotechnology and biochemical/bioprocess engineering.

Hannover, March 1994   Karl Schügerl

Contents

1 Introduction

Cultivation media for microorganisms consist of carbon-, nitrogen-, phosphate-sources and nutrient salts with trace elements and vitamins. In laboratories, synthetic media with well-defined composition can be used. In industrial production, complex media consisting of agricultural (by)products and wastes (molasses, pharma medium, cornsteep liquor, yeast extract, soybean meal, peanut flour, cottonseed meal, fish meal, etc.) with poorly defined composition are employed. During the cultivation and production process, the medium components are consumed and converted into other compounds. The aim of the downstream processing is the recovery and purification of the valuable products from this complex fermentation broth.

Primary and secondary metabolites are excreted into the cultivation broths by the microorganism. Their recovery can be performed by different techniques: crystallization (e.g., tetracycline), evaporation (e.g., ethanol) or after removal of the cells and solid particles (by means of filtration or centrifugation) by precipitation (e.g., Ca-salt of citric acid and lactic acid), fractional distillation (e.g., acetone/butanol), adsorption (e.g., cephalosporin C), ion exchange (e.g., amino acids), ultrafiltration (e.g., acetic acid), electrodialysis (e.g., lactic acid) or solvent extraction.

Solvent extraction is the most called-for method for the recovery of hydrophilic substances, and, therefore, a method for separating well-soluble metabolites from the cultivation medium. The classical extraction processes use organic solvents, which are often rarely suitable for effective recovery of the solute. Recently, new extractants were developed which form specific adducts (complexes or compounds) with a metabolite in question and allow its recovery with high efficiency and selectivity.

After considering the reaction engineering principles of extraction in Chapter 2, and extraction equipment in Chapter 3, the extraction of primary and secondary metabolites with different extractants are treated in Chapter 4. Besides solvent extraction, novel separation techniques with liquid membrane, microemulsion, and reversed micelles are also presented.

2 Reaction Engineering Principles

2.1 Definitions

2.1.1 Mass Action Law Equilibria

Equilibrium partitioning of nonelectrolytes between two immiscible phases is achieved, when the chemical potential of the solute is equal in the two phases

$$\mu_a - \mu_o, \tag{2.1}$$

where the potential in the aqueous phase is given by

$$\mu_a = \mu_a^o + RT \ln x_a \gamma_a \tag{2.2}$$

and that of the organic phase by

$$\mu_o = \mu_o^o + RT \ln x_a \gamma_o \tag{2.3}$$

In Eqs. (2.2) and (2.3), μ_a^o and μ_o^o are the standard potentials, x_a, x_o are the mol fractions and γ_a, γ_o are the activity coefficients in the aqueous and organic phases. At low mol fractions (usually below $0.1 \, \text{mol} \, l^{-1}$) the activity coefficients are unity (ideal solutions),

$$\gamma_a, \gamma_o \rightarrow 1 \tag{2.4}$$

The standard energy of the transfer of the solute is given by

$$\Delta G_x^o = \mu_o - \mu_a = -RT \ln P_x, \tag{2.5}$$

where

$$P_x = \frac{x_o}{x_a} \tag{2.6}$$

is the mol fraction based partition coefficient. The molar concentration based partition coefficient

$$P_c = \frac{c_o}{c_a}, \tag{2.7}$$

where c_o and c_a are the molar concentrations of the solute ($\text{mol} \, l^{-1}$) in the organic and aqueous phases, and the solute mols and phase masses based partition coefficient are

$$P_w = \frac{w_o}{w_a},$$ (2.8)

where w_o and w_a are the molar concentrations of the solute with regard to the mass of the phases (mol kg^{-1}) and are more frequently used than the mole fraction based partition coefficient. The relationship between these partition coefficients are

$$P_x = P_c(v_o/v_a)$$ (2.9)

$$P_w = P_c(\rho_a/\rho_o),$$ (2.10)

where v_a, v_o are the molar volumes and ρ_a, ρ_o are the densities of the aqueous and organic phases.

Since the mutual solubilities of the aqueous and organic phases are usually not negligible, the two phases have to be mutually saturated for the determination of the partition coefficients. When using the molar concentration based partition coefficient P_c, the standard free energy of the solute transfer is given by

$$\Delta G_c^\circ = -RT \ln P_c = \Delta G_x^\circ + RT \ln(v_o/v_a),$$ (2.11)

where v_o, v_a are the molar volumes of the organic and aqueous phases.

The distribution ratio D_c is often used in literature

$$D_c = \frac{c_{ot}}{c_{at}},$$ (2.12)

where c_{ot}, c_{at} (mol l^{-1}) are the total measured solute concentrations in the organic and aqueous phases without any correction.

Because of the necessary corrections, P_c and D_c are generally not identical.

Some solutes interact in this way with the solvent (extractant Ex). The stoichiometry and the equilibrium constant are

$$So_a + n\,Ex_o \overset{K_{en}}{\rightleftharpoons} (SoEx_n)_o$$ (2.13)

$$K_{en} = \frac{[SoEx_n]_o}{[So]_a[Ex]_o}$$ (2.14)

The square brackets denote equilibrium concentrations in mol l^{-1}.

The distribution ratio D_c is given by

$$D_c = \frac{[SoEx_n]_o + [So]_o}{[So]_a} = \frac{[So]_o + K_{en}[So]_o[Ex]_o^n}{[So]_a}$$

$$= P_c(1 + K_{en}[Ex]_o^n),$$ (2.15)

where

$$P_c = \frac{[So]_o}{[So]_a}.$$ (2.16)

A plot of $\log[(D_c/P_c) - 1]$ vs $\log[Ex]_o$ yields a straight line with the slope of n and an intercept of $\log K_{en}$.

Some solutes (e.g. organic acids HS) dissociate in the aqueous phase

$$HS \overset{K}{\rightleftharpoons} H^+ + S^-, \tag{2.17}$$

where H^+ is the proton and S^- the acid anion.

$$K = \frac{[H^+][S^-]}{[HS]}. \tag{2.18}$$

The partition coefficient is based on the undissociated solute

$$P_c = \frac{[HS]_o}{[HS]_a}. \tag{2.19}$$

When the solute forms dimers in the organic phase, the following relationships hold true:

$$2HS_o \overset{K_{di}}{\rightleftharpoons} (HS)_{2,o} \tag{2.20}$$

$$K_{di} = \frac{[(HS)_2]_o}{[HS]_o^2}. \tag{2.21}$$

The distribution ratio D_c is again based on the total concentrations of the solute in all its possible forms in the aqueous phase C_{HAat} and in the organic phase C_{HAot}:

$$D_c = \frac{C_{HSot}}{C_{HSat}} = \frac{[HS]_o + 2[(HS)_2]_o}{[HS]_a + [S^-]_a} = \frac{[HS]_a P_c + 2K_{di}[HS]_o^2}{[HS]_a + K[HS]_a/[H^+]_a}$$

$$= \frac{[HS]_a P_c + 2K_{di} P_c^2 [HS]_a^2}{[HS]_a(1 + K/[H^+]_a)} = \frac{P_c + 2P_c^2 K_{di}[HS]_a}{1 + K/[H^+]_a}. \tag{2.22}$$

The Mass Action Law Equilibria for the extraction of monocarboxylic acid by strong solvating extractants, such as organo-phosphorous compounds, are strongly influenced by the dissociation of the acid in the aqueous solution

$$HS \overset{K}{\rightleftharpoons} H^+ + S^-, \tag{2.23}$$

$$K = \frac{[H^+][S^-]}{[HS]} \tag{2.24}$$

and by the formation of the acid solvating extractant Ex in the organic phase

$$HS_a + n Ex_o \overset{K_{en}}{\rightleftharpoons} HSEx_{no}, \tag{2.25}$$

$$K_{en} = \frac{[HSEx_n]_o}{[HS]_a [Ex]_o^n}. \tag{2.26}$$

The distribution ratio D_c on the molar concentration basis is given by

$$D_c = \frac{C_{HSot}}{C_{HSat}} = \frac{[HSEx_n]_o}{[HS]_a + [S^-]_a} = \frac{K_{en}[HS]_a[Ex]_o^n}{[HS]_a + K[HS]_a/[H^+]_a}$$

$$= \frac{K_{en}[Ex]_o}{1 + K/[H^+]_o}, \tag{2.27}$$

and the concentration of the unbound extractant $[Ex]_o$ in the organic phase

$$[Ex]_o = [Ex_t]_o - n[HS]_o, \tag{2.28}$$

where $[Ex_t]_o$ is the overall concentration of the extractant. If the ratio $K/[H^+]_a$ is constant or has a negligible small value, the solvation number n can be obtained by the differentiation of D_c with respect to $[Ex]_o$.

A plot of

$$\log D_c = \log K_{en} + n \log[Ex]_o \tag{2.29}$$

yields the slope n and the intercept K_{en}.

Equations (2.25) and (2.26) neither take the free acid nor the dimer into account, since in the presence of strongly solvating agents no unsolvated acid monomer and dimer are present.

Secondary, tertiary or quarternary amines with long hydrocarbon chains – to make them water immiscible and to reduce their loss – are often used as extractants.

The Mass Action Law Equilibria between the proton bearing organic compound HS in the aqueous phase and the long chain aliphatic amine extractant A in the organic phase is given by

$$HS_a + A_o \overset{K_e}{\rightleftharpoons} HSA_o, \tag{2.30}$$

$$K_e = \frac{[HSA]_o}{[HS]_a[A]_o}. \tag{2.31}$$

In some cases this acid-base type reaction is not stoichiometric. If the extractant takes acid in excess, the following relationships are valid:

$$AHS_o + m\,HS_a \overset{K_{em}}{\rightleftharpoons} AHS(HS)_{mo}, \tag{2.32}$$

$$K_{em} = \frac{[AHS(HS)_m]_o}{[AHS]_o[HS]_o^m}. \tag{2.33}$$

In some publications, the equilibrium distribution coefficient K_D is used, which, defined as the weight fraction of the solvent phase per weight fraction in the aqueous phase, is determined empirically. The relationship between K_D and

D_c is given by

$$D_c = K_D \frac{\rho_o}{\rho_a}, \tag{2.34}$$

where ρ_o, ρ_a are the densities of the organic and aqueous phases.

The degree of extraction E is often used in practice. It is the ratio of the amount of the solute extracted into the solvent (e.g. organic) phase to its amount in the original (e.g. aqueous) phase in the equilibrium.

$$E = \frac{[So]_o V_o}{[So] V_o + [So] V_s} = \frac{P_c}{P_c + r}, \tag{2.35}$$

where $r = V_a/V_o$. V_a, V_o are volumes of the aqueous and organic phase, P_c is the partition coefficient,

$$P_c = \frac{[So]_o}{[So]_a}. \tag{2.36}$$

Liquid-liquid equilibrium data can be found in [1–11].

2.1.2 Phase Diagrams

Beside the mass action law equilibria the second important approach is the phase rule [12]

$$F = N - \phi + 2, \tag{2.37}$$

where F is the number of degrees of freedom or the number of independent variables (T, p, c), which have to be fixed to define completely a system at

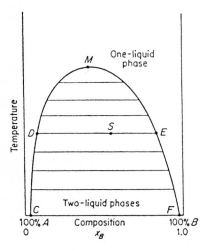

Fig. 2.1. Example of a phase diagram with temperature and concentration as independent variables and with an upper critical solution temperature [10]

equilibrium. N is the number of components or the lowest number of independently variable components required to express the composition of each phase and ϕ is the number of phases. In extraction $\phi = 2$, that is $F = N$.

In the simplest case we have three components: liquid A, from which the solute is removed, liquid B, to which the solute is transferred, and the solute. The prerequisite of the extraction is the presence of the two phases: the raffinate (solvent A + solute) and extract (solvent B + solute) phases.

First, the two-component systems (liquid A and liquid B) should be considered. If the two-component systems are completely miscible, no extraction is possible. All practical extraction systems are partially miscible. Since

$$N = 2, \quad \phi = 2, \quad F = 2 - 2 + 2 = 2,$$

the system is bivariant, and the variable temperature, pressure and concentration may be independently varied within limits in pairs without changing the number of phases.

The extraction is usually performed at a constant pressure. Figure 2.1 shows a phase diagram with temperature and concentration as independent variables in a special case, i.e., with an upper critical solution temperature. In this diagram, the equilibrium vapor pressure of the liquid is constant.

Curve CDM shows the composition of the saturated solutions of B in A and FEM shows A in B as a function of the temperature. Above curve CDMEF, a single phase, and below it, two mutually saturated phases prevail.

Point S represents two saturated solutions with the composition of D and E. Any mixture represented by a point on line DE forms the same two saturated solutions D and E, at equilibrium, but the relative amounts of these solutions depend on their position on this tie-line and can be calculated from the material balance, e.g., for the mixture at S:

$$S = D + E,$$

where S, D and E are the weights.

A material balance for component B is given:

$$Sx_{BS} = Dx_{BD} + Ex_{BE},$$

$$\frac{D}{E} = \frac{x_{BE} - x_{BS}}{x_{BS} - x_{BD}} = \frac{ES}{SD},$$

where x_{BS}, x_{BD} and x_{BE} are weight fractions of B. This means that the relative weights of the two saturated phases are inversely proportional to the lengths of the line segments.

In the case of three-component systems, the equilateral triangle is often used for representing the system compositions (Fig. 2.2). As in the case of binary systems, the pressure effect is relatively insignificant and therefore the vapor pressure is not taken into consideration for the three-component system either.

In this diagram, the sum of the perpendiculars from any point within the triangle to the three sides is equal to the altitude. The length of the altitude is

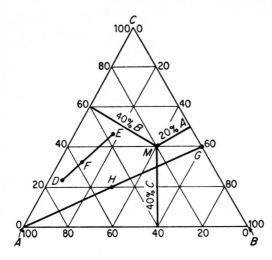

Fig. 2.2. Triangular diagram for the three component systems [10]

then allowed to represent 100% composition, and the length of the perpendiculars from any point the percentages of the components. The apexes of the triangle represent the pure components A, B and C, respectively. Any point on the side of the triangle represents a binary mixture of the two components at the ends of the side.

When mixing D and E, the resulting mixture will lie on the tie-line DE. When we mix E and D in such a way that the resulting mixture should have the composition F, then the weight ratio of E to D is given by

$$\frac{E}{D} = \frac{FD}{EF} \,.$$

For extractions only the three component systems are interesting which form at least one pair of partially miscible liquids. This system can be represented by Fig. 2.3 [10]. At constant temperature and pressure, components A and C, as well as B and C are miscible in all portions. A and B are partially miscible, and points D and E represent the saturated solutions in the binary system.

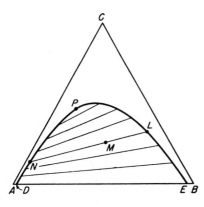

Fig. 2.3. Ternary liquid equilibria in a system with one pair of partially miscible liquids [10]

In our case, components are A (water), B (organic solvent), C (solute), D (water-saturated by the organic phase) and E (organic phase saturated by water). All points above the binodal curve DNPLE represent homogeneous mixtures which are not of interest for extraction. Below the binodal curve, two phases are present. Every mixture represented by a point in this field decomposes into two phases with a definite composition. E.g., the mixture correspondence point M decomposes into mixtures N and L. All points that lie on line NL yield the same two mixtures according to points N and L. Only the mass fractions vary according to the ratio of the segments

$$\frac{L}{N} = \frac{MN}{ML}.$$

When increasing the concentration of C, the mutual solubility of A and B increases; at plait point P the two branches of the solubility curve merge and the tie-lines vanish. When the temperature and the pressure are kept constant,

$$F = N - \phi,$$

and in the case of a three-component system

$F = 3 - 1 = 2$ for the mixture of a one-liquid phase,

$F = 3 - 2 = 1$ for mixtures of two-liquid phases and

$F = 0$ at the plait point.

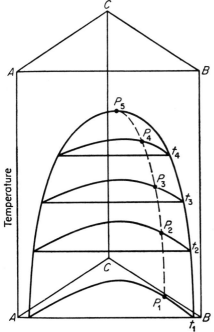

Fig. 2.4. Example of the temperature influence of the plait point [10]

If the temperature is varied the composition of the plait point P mixture changes. Figure 2.4 shows an example for the temperature effect on the plait point. (For four components systems, see Treybal [10]).

2.1.3 Diffusion and Mass Transfer

The molar flux of the solute j_S across the surface area A in the case of unidirectional diffusion in a liquid in rest is given by the first Fick's law [13]:

$$j_S = - D_S \frac{dc_S}{dz}, \tag{2.38a}$$

where D_S is the diffusivity of the solute in the liquid, c_S is the molar concentration of the solute, and z is the coordinate normal to the surface A.

For a volume of the fluid of the unit cross-section and dz long, the rate of change of S in the volume is given by Fick's second law:

$$\frac{\delta c_S}{\delta t} = - D_S \frac{\delta^2 c_S}{\delta z^2}. \tag{2.38b}$$

For the calculation and experimental determination of D_S, see e.g. [16].

The molar flux of the solute through the interface from the aqueous to the organic phase – both of them in motion – is given according to the two-film theory of Whitman [14]:

$$- j_S = k_{Sa}(c_{Sa} - c_{Sa}^i) = k_{So}(c_{So}^i - c_{So}), \tag{2.39}$$

where k_{Sa}, k_{So} are the particular mass transfer coefficients of the solute in the stagnant films of the aqueous and organic phases at their interface, c_{Sa}, c_{So} are solute concentrations in the bulk of the aqueous and organic phases, and c_{Sa}^i, c_{So}^i are solute concentrations in the aqueous and organic phases at the interface.

According to Whitman there is an equilibrium at interface i:

$$P_c = \frac{c_{So}^i}{c_S^i}, \tag{2.40}$$

and in the stagnant films with the thickness δ_o and δ_a, mass transfer occurs only by molecular diffusion:

$$k_{Sa} = \frac{D_{Sa}}{\delta_a} \quad \text{and} \quad k_{So} = \frac{D_{So}}{\delta_o} \tag{2.41}$$

(For alternative models, see Chapter 2.2).

Since c_{Sa}^i and c_{So}^i cannot be determined experimentally, it is necessary to eliminate them from Eq. (2.39).

If Eq. (2.41) holds true and the mass transfer resistances in the two phases are added, Eq. (2.42) is valid:

$$\frac{1}{K} = \frac{1}{k_{Sa}} + \frac{1}{k_{So}}, \tag{2.42}$$

where K is the overall mass transfer coefficient. The solute mass transfer rate N_S from the bulk water phase to the bulk organic phase can be written as:

$$N_S = K_{So}(c_{So}^* - c_{So}) \tag{2.43a}$$

or

$$N_S = K_{Sa}(c_{Sa} - c_{Sa}^*) . \tag{2.43b}$$

Here K_{Sa} and K_{So} are the overall mass transfer coefficients with regard to the aqueous and organic phases, c_{Sa} and c_{So} are the bulk concentrations in the aqueous and organic phases, and c_{Sa}^* and c_{So}^* are the hypothetical solute concentrations in the aqueous and organic phases (Fig. 2.5).

As one can see from Fig. 2.5, the two resistances are replaced by a single resistance in one of the phases and the two driving forces:

$$(c_{Sa} - c_{Sa}^i) = c_{Sa} \quad \text{and} \quad (c_{So}^i - c_{So}) = c_{So}$$

by one of the hypothetic driving forces:

$$(c_{Sa} - c_{Sa}^*) = c_{Sa}^* \quad \text{or} \quad (c_{So}^* - c_{So}) = c_{So}^* .$$

c_{So} concentration of S in the organic phase
c_{Sa} concentration of S in the aqueous phase

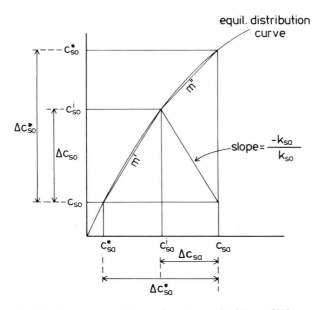

Fig. 2.5. Explanation of the two-film theory of Whitman [10]

According to Fig. 2.5:

$$c_{So}^* = c_{So} + m'' c_{Sa},$$

$$\frac{N_S}{K_{So}} = \frac{N_S}{k_{So}} + \frac{m'' N_S}{k_{Sa}},$$

$$\frac{1}{K_{So}} = \frac{1}{k_{So}} + \frac{m''}{k_{Sa}}, \tag{2.44a}$$

$$c_{Sa}^* = c_{Sa} + \frac{c_{So}}{m'},$$

and

$$\frac{1}{K_{Sa}} = \frac{1}{k_{Sa}} + \frac{1}{m' k_{So}}. \tag{2.44b}$$

If the equilibrium distribution of the solute strongly favors e.g the organic phase, m' will be very large, $K_a \approx k_{Sa}$ and the principal resistance to mass transfer lies in the aqueous phase. In Fig. 2.5, $c_{Sa}^* \approx c_{Sa}$, $c_{Sa}^* \sim c_{Sa}^i$, $c_o \approx c_{So}^i$ and $P_c \approx c_{So}/c_{Sa}^*$ hold true. If the distributed solute favors the aqueous phase at equilibrium, $K_{So} \approx k_{So}$, and the resistance in the organic phase controls the mass transfer, $c_{So}^* \approx c_{So}$, $c_{So}^* \approx c_{So}^i$, $c_{Sa} \approx c_{Sa}^i$ and $P_c \approx c_{So}^*/c_{Sa}$ hold true.

Substituting c_{So}^* and c_{Sa}^* in Eqs. (2.43a) and (2.43b) by the corresponding relationships,

$$N_S \approx K_{So}(P_c c_{Sa} - c_{So}) \tag{2.45a}$$

and

$$N_S \approx K_{Sa}\left(c_{Sa} - \frac{c_{So}}{P_c}\right) \tag{2.45b}$$

are obtained.

2.1.4 Stagewise Counter-Current Multiple Contact with a Single Solvent

This is the most common separation of the components of a solution by extraction. The solution to be separated, usually called feed (F), and the extraction pure solvent (S) are mixed in a so-called stage and allowed to approach equilibrium, and then the two immiscible phases are separated and withdrawn.

The solute-rich phase leaving the stage is called 'extract (E)' and the solute 'lean phase raffinate (R)' (Fig. 2.6).

For multiple continuous contact, a cascade of stages are used with feed (F) and pure solvent (S) entering at opposite ends of the cascade, and extract and raffinate are also withdrawn at the opposite ends of the cascade (Fig. 2.7).

When using a triangular diagram for ternary systems and assuming that the locations of F, E_1, R_n, and S are known, the material balance for the cascade is

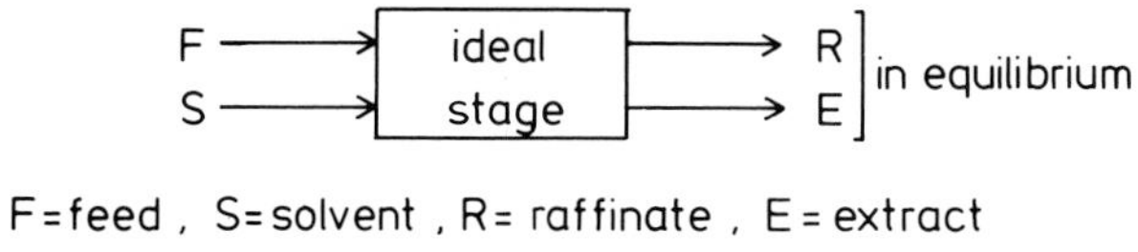

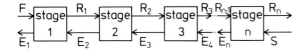

F = feed , S = solvent , R = raffinate , E = extract

Fig. 2.6. Explanation of the definition of the ideal stage

Fig. 2.7. Flowsheet for countercurrent continuous multiple contact (multistage) extraction. F = feed, S = solvent, R = raffinate, E = extract

given by

$$F + S = E_1 + R_n = M \tag{2.46a}$$

or

$$F - E_1 = R_n - S = O. \tag{2.46b}$$

Point O is the operating point. It can be located by extending lines E_1F, and SR_n to the intersection (Fig. 2.8).

The material balances of stages 1 through m are given [10]:

$$F + E_{m+1} = E_1 + R_m \tag{2.47a}$$

or

$$F - E_1 = R_m - E_{m+1} = O \tag{2.47b}$$

For the mth stage:

$$R_{m-1} + E_{m+1} = R_m + E_m \tag{2.48a}$$

or

$$R_{m-1} - E_m = R_m - E_{m+1} = O \tag{2.48b}$$

Since R_m and E_m are in equilibrium, they are on the tie-line. On the other hand, Eqs. (2.47a) and (2.47b) indicate that any extract E_{m+1} can be located from any raffinate R_m by extending the line OR_m to the B-rich binodal curve (Fig. 2.8).

Operating point O can be located by means of Eq. (2.46b) and is determined by starting with F and E_1 as well as R_n and S. E_1 is in equilibrium with R_1, and thus the crossing point of the tie-line with the binodal curve on the A-rich side determines R_1. The crossing point of the operating line OR_1 with the binodal curve on the B-rich side determines E_2, and so on. For extension and limitation of the number calculations of ideal stages, see the book of Treybal [10].

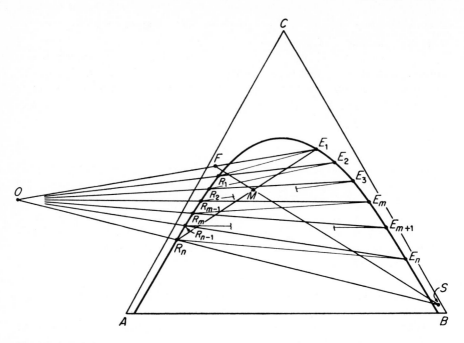

Fig. 2.8. Calculation of the stages for a countercurrent multistage extractor [10]

The numbers of the theoretical stages calculated in this way are not identical with the number of real stages, because the equilibrium cannot be attained in the stages. The ratio of the number of theoretical stages to the real stages, the stage efficiency, has to be taken into account (see [10]).

2.1.5 System Models

Extraction columns are often used that do not have real stages, or for which there is no well-defined relationship between real and theoretical stages. In these cases models can be used which were developed for chemical reactors [15].

The main idea of these models is that in a tubular reactor the axial dispersion, which superposes the convection, influences the conversion in the reactor or the degree of extraction in the extraction column.

Two types of models are used to take into account the influence of the axial dispersion: the cascade and the dispersion models. The cascade models assume that the column consists of ideally mixed discrete stages of equal size. One stage $(N = 1)$ corresponds to an ideal mixer unit. The diminishing influence of the axial dispersion is described with an increasing number of ideal stages N. The absence of axial dispersion is represented by the upper limit $N = \infty$ and the ideally mixed column by a single stage $(N = 1)$. The dispersion model assumes that the liquid flow in the column is superposed by axial dispersion with the dispersion coefficient D_{ax}.

The cascade model assumes no interaction of the active section of the column with its upstream and downstream sections. The advantage of this model is that the solute balance can be described by a simple algebraic equation system. In a single nth stage of the cascade the solute balance in the continuous phase is given by:

$$F_{an-1} c_{San-1} - F_{an} c_{San} = K_{San} a_{an} (c_{San} - c_{San}^*) V_{an}, \tag{2.49}$$

where F_{an-1} is the aqueous phase fed into stage n, F_{an} is the aqueous phase withdrawn from stage n, c_{San-1} is the solute concentration in the aqueous feed into stage n, and c_{San} is the solute concentration in the aqueous phase withdrawn from stage n, K_{San} is the overall mass transfer coefficient of solute from phase a to phase o with regard to phase a in stage n, a_{an} is the specific liquid/liquid interfacial area with regard to the volume of the aqueous phase in stage n, V_{an} is the volume of the aqueous phase in stage n, c_{an} is the bulk concentration of solute in the aqueous phase in stage n, and c_{an}^* is the hypothetical solute concentration in the aqueous phase in equilibrium with its concentration in the bulk of the organic phase c_{San} in stage n.

When $F_{an-1} \approx F_{an}$, the throughput of the aqueous phase in stage n is given by F_{an}. Dividing Eq. (2.49) by V_{an},

$$\tau (c_{San-1} - c_{San}) = K_{San} a_{an} (C_{San} - c_{San}^*), \tag{2.50}$$

where $\tau_{an} = F_{an}/V_{an}$, the mean residence time of the aqueous phase in stage n, is obtained. Starting with $n = 1$, the balances are calculated for each of the stages up to N. The model parameters τ and N can be determined by means of the distribution of residence times of the phase a (see Chapter 2.4.3).

The mass balance of the solute in the dispersion model is described by a partial differential equation. For steady state operation, the solute balance in the continuous aqueous phase is given by

$$u_a \frac{\delta c_{Sa}}{\delta z} - D_{axa} \varepsilon_a \frac{\delta c_{Sa}^2}{\delta z^2} = K_{Sa} a_a (c_{Sa} - c_{Sa}^*), \tag{2.51}$$

where u_a is the superficial velocity of the aqueous phase, D_{axa} is the axial dispersion coefficient in the continuous aqueous phase, c_{Sa} is the solute concentration in the aqueous phase, z is the longitudinal coordinate, ε_a is the holdup of the aqueous phase, K_{Sa} is the overall mass transfer coefficient of the solute with regard to the aqueous phase, c_{Sa}^* is the hypothetical solute concentration in the aqueous phase in equilibrium with the solute concentration in the bulk of the organic phase, and a_a is the specific aqueous/organic phase interfacial area with regard to the volume of the aqueous phase.

Contrary to the cascade model, the dispersion model can take into account the interaction of the active section of the column with its inactive upstream and downstream sections by means of the boundary conditions. The boundary conditions depend on the extractor construction.

The feeding and withdrawal connections of the aqueous phase to the column usually have much smaller cross-sections than those of the column. Therefore, the system boundaries with regard to the dispersion in the aqueous phase can be considered as being closed. Thus, the cascade model as well as the dispersion model, the latter usually with "closed" boundaries with regard to the dispersion term at the entrance and the exit of the aqueous phase, can be used:

$$(u_a)_1 (c_{Sa})_1 = u_a c_{Sa} - D_{axa} \varepsilon_a \frac{dc_{Sa}}{dz} \quad \text{at } z = 0 \qquad (2.51a)$$

entrance of the aqueous phase

and

$$u_a c_{Sa} - D_{axa} \varepsilon_a \frac{dc_{Sa}}{dz} = (u_a)_2 (c_{Sa})_2 \quad \text{at } z = L \qquad (2.51b)$$

exit of the aqueous phase,

where the indices 1 and 2 mean the entrance and the exit boundaries of the aqueous phase.

The parameters of the dispersion model (Eq. 2.51) are: $\tau_a = L/u_a^*$, the mean residence time of the aqueous phase and D_{axa} is the dispersion coefficient in the aqueous phase; $u_a^* = u_a/\varepsilon_a$, the true liquid velocity.

For the determination of the model parameters, see Chapter 2.4.3.

2.1.6 Symbols for Sect. 2.1

M mass, L length, T time, θ temperature, 1 dimensionless

Sect. 2.1.1

A	amine extractant	
c_a	molar concentration of solute in the aqueous phase	$M L^{-3}$
c_{at}	total measured solute concentration in the aqueous phase	$M L^{-3}$
c_o	molar concentration of the solute in the organic phase	$M L^{-3}$
c_{ot}	total measured solute concentration in the organic phase	$M L^{-3}$
D_c	distribution ratio	1
E	degree of extraction	1
Ex	organophosphorus extractant	
ΔG_x^o	standard energy of solute transfer	$M L^2 T^{-2}$
H^+	proton	

HS	acid	
$(HS)_2$	acid dimer	
HSA	acid extractant complex Eq. (2.30)	
HSEx	acid extractant complex Eq. (2.25)	
K	dissociation constant Eq. (2.18)	1
K_{di}	dimerisation equilibrium constant Eq. (2.21)	1
K_e	equilibrium constant Eq. (2.31)	1
K_{em}	equilibrium constant Eq. (2.33)	1
K_{en}	equilibrium constant Eq. (2.14)	
P_c	molar concentration based partition coefficient	1
P_x	mole fraction based partition coefficient	1
P_w	partition coefficient based on w_a and w_o	1
R	gas constant	$L^2 T^{-2} \theta^{-1}$
S^-	acid anion	
So	solute	
T	absolute temperature	θ
x_a	mole fraction of the solute in the aqueous phase	1
x_o	mole fraction of the solute in the organic phase	1
V_a	volume of the aqueous phase	L^3
v_a	molar volume of the aqueous phase	$L^3 M^{-1}$
V_o	volume of the organic phase	L^3
v_o	molar volume of the organic phase	$L^3 M^{-1}$
w_a	molar concentration of the solute in the aqueous phase with regard to the mass of the aqueous phase	1
w_o	molar concentration of solute in the organic phase with regard to the mass of the organic phase	1
γ_a	activity coefficient of the solute in the aqueous phase	1
γ_o	activity coefficient of the solute in the organic phase	1
ρ_a	density of the aqueous phase	$M L^{-3}$
ρ_o	density of the organic phase	$M L^{-3}$
μ_a	chemical potential of solute in the aqueous phase	1
μ_a^o	standard chemical potential of the solute in the aqueous phase	1
μ_o	chemical potential of the solute in the organic phase	1
μ_o^o	standard chemical potential of the solute in the organic phase	1

Sect. 2.1.2

A, B, C	represent the pure components in Fig. 2.2	
E, D, S	represent particular mixtures in Fig. 2.2	
F	number of degrees of freedom in the phase rule	
N	number of components in the phase rule	
x_{BS}, x_{BD}, x_{BE}	weight fractions of B in S, D and E	
ϕ	number of phases in phase rule	

Sect. 2.1.3

c_S	molar concentration of the solute	$M L^{-3}$
c_{Sa}, c_{So}	molar concentrations of the solute in the aqueous and organic bulk phases	$M L^{-3}$
c_{Sa}^i, c_{So}^i	molar concentrations of the solute in the aqueous and organic phases at the interface	$M L^{-3}$
c_{Sa}^*, c_{So}^*	hypothetical molar concentrations of the solute in the aqueous and organic phases (Fig. 2.5)	$M L^{-3}$
D_S	diffusivity of solute	$L^2 T^{-1}$
j_S	solute flux	$M L^{-2} T^{-1}$
k_{Sa}, k_{So}	particular mass transfer coefficients of the solute in the aqueous and organic phases	$L T^{-1}$
m', m''	slopes in Fig. 2.5	1
N_S	solute mass transfer rate	$M L^{-2} T^{-1}$
t	time	T
z	coordinate normal to the surface	L

Sect. 2.1.4

E	extract
F	feed
M	operating point Eq. (2.46a)
O	operating point Eq. (2.46b)
R	raffinate
S	solvent

Sect. 2.1.5

a_a	specific aqueous/organic phase interfacial area with regard to the volume of the aqueous phase	L^{-1}
a_{an}	specific aqueous/organic phase interfacial area with regard to the volume of the aqueous phase in stage n	L^{-1}

c_{Sa}	solute concentration in the aqueous phase	$M L^{-3}$
c_{Sa}^*	hypothetical solute concentration in the aqueous phase in equilibrium with its concentration in the bulk of the organic phase	$M L^{-3}$
c_{San}	solute concentration in the aqueous phase in stage n	$M L^{-3}$
c_{San}^*	hypothetical solute concentration in the aqueous phase in equilibrium with its concentration in the bulk of the organic phase in stage n	$M L^{-3}$
D_{axa}	axial dispersion coefficient in the continuous aqueous phase	$L^2 T^{-1}$
F_{an}	throughput of the aqueous phase in stage n	$L^3 T^{-1}$
K_{Sa}	overall mass transfer coefficient of the solute with regard to the aqueous phase	$L^2 T^{-1}$
K_{San}	overall mass transfer coefficient of the solute with regard to the aqueous phase in stage n	$L^2 T^{-1}$
N_a	number of ideally mixed stages for the aqueous phase	1
u_a	superficial velocity of the aqueous phase	$L T^{-1}$
V_{an}	volume of the aqueous phase in the stage n	L^3
z	longitudinal coordinate	L
ε_a	holdup of the aqueous phase	1
τ_{an}	mean residence time of the aqueous phase in stage n	T

2.2 Mechanism of Mass Transfer Between Two Phases

In general, mass transfer resistance of the interface is neglected in contrast to those in the phases near to the interface. Therefore, only the resistances in the phases will be taken into account.

The two major models of the mass transfer mechanism between the two phases are the film theory [17] and the penetration theory [18–20]. The film theory assumes that there is a region (film), in which steady-state molecular transfer is rate-controlling. The penetration theory assumes that the interface is continuously replaced by eddies, and that unsteady-state molecular transfer between the eddies and their environment controls the transfer in this region. Toor and Marcello [21] have shown that the validity of penetration and film models depend on the age of the fluid elements at the interface of the phases: the transfer to young elements at the interface follows the penetration theory, and the transfer into old elements follows the film theory. The transfer equation in one of the phases is given by Eq. 2.52

$$\frac{\delta C}{\delta t} = - D \frac{\delta^2 C}{\delta y^2} \tag{2.52}$$

This transfer equation neglects the velocity gradients at the interface. The boundary conditions are:

$$t = 0 \qquad C = C_L \tag{2.53}$$

$$y = 0 \qquad C = C_i \tag{2.54}$$

$$y = L \qquad C = C_L \tag{2.55}$$

where D is the diffusivity of the solute, t the time of the mass transfer, C the concentration of the solute, y the distance from the interface, C_i the C at the interface $y = 0$, C_L the C at $y = L$, and L the thickness of the region in which the molecular mass transfer controls.

The solutions of Eqs. (2.52) to (2.55) are [21, 22]: for a short period of time $(t < L^2/D)$

$$N = \Delta C \sqrt{\frac{D}{\pi t}} \left[1 + 2 \sum_{n=1}^{\infty} \exp \left\{ -\frac{n^2 L^2}{Dt} \right\} \right] \tag{2.56a}$$

and for a longer period of time $(t > L^2/D)$:

$$N = \Delta C \frac{D}{L} \left[1 + 2 \sum_{n=1}^{\infty} \exp \left\{ -n^2 \pi^2 \frac{Dt}{L^2} \right\} \right] \tag{2.56b}$$

as the series in Eq. (2.56a) rapidly converges for a short period of time and in Eq. (2.56b) for a longer period of time. Hence for short periods:

$$N = \Delta C \sqrt{\frac{D}{\pi t}} \tag{2.57}$$

i.e., the penetration theory, and for longer times:

$$N = \Delta C \frac{D}{L} \tag{2.58}$$

i.e., the film theory is approached.

In Eqs. (2.56) to (2.58) are N the point mass transfer rate and $\Delta C = C_i - C_L$, the driving force of the mass transfer.

Higbie [18] assumed a deterministic contact time t' of the eddies at the interface. This assumption holds true for the mean mass transfer rate N for short time:

$$N = \Delta C \sqrt{\frac{D}{\pi t'}} \left[1 + 2\sqrt{\pi} \sum_{n=1}^{\infty} \text{ierfc} \frac{nL}{\sqrt{Dt'}} \right] \tag{2.59a}$$

For a short time $(t' < L^2/D)$, the equation is reduced to the Higbie-Model:

$$N = \Delta C \sqrt{\frac{D}{\pi t'}} \tag{2.60}$$

Danckwerts assumed that the elements of the surface area interchanged in a random fashion within an average contact time of 1/S. Hence, for a short average contact time:

$$N = 2\Delta C \sqrt{DS} \left[1 + 2 \sum_{n=1}^{\infty} \exp\left\{ - 2nL \sqrt{\frac{S}{D}} \right\} \right] \qquad (2.61a)$$

For a short contact time, Eq. (2.61a) is reduced to the Danckwerts model:

$$N = 2\Delta C \sqrt{DS} \qquad (2.62)$$

For a long contact time the assumption of Higbie yields:

$$N = \Delta C \frac{D}{L} \left[1 + \frac{2}{\pi^2} \frac{L^2}{Dt'} \left(\frac{\pi^2}{6} - \sum_{n=1}^{\infty} \exp\left\{ - n^2\pi^2 \frac{Dt'}{L^2} \right\} \right) \right] \qquad (2.59b)$$

and that of Danckwerts leads to:

$$N = \Delta C \frac{D}{L} \left[1 + 2 \sum_{n=1}^{\infty} \frac{1}{1 + n^2\pi^2 D/SL^2} \right] \qquad (2.61b)$$

For long contact times ($t' > L^2/D$), Eqs. (2.59b) and (2.61b) approach the film model:

$$N = \Delta C \frac{D}{L} \qquad (2.63)$$

At low Schmidt numbers, $Sc = v/D$, the steady state concentration gradient is formed very rapidly in any new interface element, so that, unless the rate of renewal is high enough to remove a large fraction of the interface elements before they are penetrated, most of the interface is old. Steady-state transfer takes place thoroughly and the film model holds [21].

As the Schmidt number increases, the time necessary to set up the steady-state concentration gradient increases rapidly, and even low rates of interface renewal are sufficient to keep most elements from being penetrated. The transfer then follows the penetration theory and the transfer rate is a function of the rate of interface renewal [21].

When the conditions are such that the interface contains young and old elements as well as middle aged ones, the transfer characteristics are intermediate between the film and the penetration typed, and then the film penetration model holds true [21].

2.2.1 Symbols for Sect. 2.2

C	concentration of solute	ML^{-3}
C_i	C at the interface	ML^{-3}
C_L	C at y = L	ML^{-3}
ΔC	= $C_i - C_L$ driving force of mass transfer	ML^{-3}
D	diffusivity of solute	$L^2 T^{-1}$

L	thickness of the boundary layer at the interface in which molecular mass transfer predominates	L
N	mass transfer rate (mass flux)	$M L^{-2} T^{-1}$
Sc	$= v/D$ Schmidt number	1
s^{-1}	average contact time of the eddies at the interface according to Danckwerts	T
t	time of mass transfer	
t'	deterministic contact time of eddies at the interface according to Higbie	T
y	distance from the interface	L
v	kinematic viscosity	$L^2 T^{-1}$

2.3 Mass Transfer During Extraction

The extraction rate depends on three factors: on the interfacial area of the phases, the effective driving forces and the mass transfer coefficients in both phases.

In order to increase the interfacial area, one of the phases is usually dispersed into the other one. This dispersion process is considered in Chapter 3 (Extraction Equipment). The effective driving forces depend on the process variables, especially on the solute concentrations in the feed of the solute-containing (raffinate) phase and in the solvent phase as well as in their outlet streams.

The mass transfer takes place from the bulk of the raffinate phase to the interface and from the interface to the bulk of the solvent phase. The mass transfer rates are differently influenced by the process variables in the continuous and dispersed phases. Therefore, these two mass transfer processes have to be considered separately.

2.3.1 Mass Transfer Within the Droplet

Hydrodynamics of droplets in fluid media is considered in detail in Chapter VIII of the book by Levich [26]. Two cases can be distinguished concerning low Reynolds numbers ($Re_d \ll 1$): a) pure systems and b) systems with surface-active substances.

Figure 2.9 shows the internal movement in droplets with mobile surface, which is characteristic for pure systems. According to Rose and Kintner [33], these streamlines prevail up to $Re_d = 200$.

Pure systems have already been considered by Hadamard [27] and Rybczynski [28], who developed the well known relation for the relative drop velocity U:

$$U = \frac{2(\rho_d - \rho_c)ga^2}{3\mu_c} \frac{\mu_c + \mu_d}{2\mu_c + 3\mu_d} \tag{2.64}$$

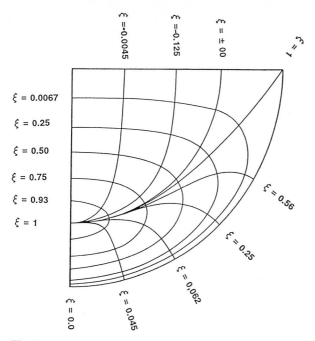

Fig. 2.9. Axial symmetric stream lines and their orthogonal trajectories in a vertical plane through the axis of a falling droplet with mobile surface at small Reynolds numbers ($Re_d < 1$) [24]

Here ρ_d and ρ_c are the density of droplet and continuous phases, g is the acceleration of gravity, a is the radius of the spherical droplet and μ_d, μ_c are the dynamic viscosity of droplet and continuous phases.

The behavior of small droplets in the presence of surface active substances was also considered by Levich [26]. These substances are adsorbed at the interface of the two liquids. Because of the movement of the interface, the interface is continuously renewed on the upstream side of the droplet. The liquid motion carries the surfactant molecules on the surface of the drop towards the rear of the drop. The resulting accumulation of surface-active substances lowers the surface tension in the rear part of the drop. The force caused by this gradient of surface tension retards the surface motion, thereby inhibiting the additional accumulation of surfactants in the back part of the drop.

Depending on the rate-determining step (adsorption, bulk diffusion, or surface diffusion) different relative velocities were evaluated:

$$U = 3U_o \frac{\mu_c + \mu_d + \gamma_n}{2\mu_c + 3\mu_d + 3\gamma_n} \tag{2.65}$$

Here

$$U_o = \frac{2(\rho_d - \rho_c)ga^2}{9\mu_c}, \tag{2.66}$$

the Stokes velocity for rigid spheres.

$\gamma_1, \gamma_2, \gamma_3$ are the retardation coefficients for adsorption, diffusion or surface diffusion as rate determining step [26]. If the retardation coefficient is large, the drop behaves like a rigid sphere according to the Stokes law.

Alternative mechanisms were proposed by Boussinesq [29] who introduced an interfacial viscosity and Savic [30], Griffith [56], as well as Davis and Acrivos [57], who assumed the formation of an inelastic film (stagnant cup) covering the rear of the droplet, whose extent is limited by the maximum surface concentration which the film can sustain without collapsing. A combination of the Frumkin-Levich model with the stagnant cup model was given by Saville [58], who investigated droplets with different diameters (Peclet numbers). Small droplets behave like a rigid sphere. With increasing droplet radius, that portion of the droplet volume increases in which a circulation movement prevails. If the droplet is large enough, the surfactant film, which forms the cup at the rear of the droplet, is destroyed by the high tension due to the high interfacial tension gradient. Thus, the interface becomes free.

Small droplets have a nearly spherical shape, and they behave in real systems like rigid bodies. Large droplets have shapes which deviate considerably from the sphere and they have a freely circulating surface. Droplets of an intermediate size are deformed and circulate to same extent.

In the bulk of small droplets, mass transfer only occurs with molecular transport. Therefore, the model of Newman [23] can be used (Table 2.1).

If the droplet is large enough, its interface is circulating freely in real systems also (e.g., in fermentation media). As long as laminar circulation in the droplet exists (at low Reynolds numbers), the model of Kronig and Brink [24] can be applied (Table 2.1).

With high Reynolds numbers in the droplet bulk, random motions due to vibrations and periodical change of the droplet shape (droplet oscillations) occur, which significantly increase the mass transfer. Mass transfer in droplets with overlapping circulation and random motion was modelled by Handlos and Baron [25] (Table 2.1).

In several papers, the Handlos-Baron model was criticized and modified. Brunson and Wellek [31] compared some of these models [24, 25, 31–35] (Table

Table 2.1. Models for mass transport from droplet to continuous phase

$\dfrac{c_0 - c_{fin}}{c_0 - c_{eq}} = 1 - \dfrac{6}{\pi^2} \sum\limits_{n=1}^{\infty} \dfrac{1}{n^2} \exp\left[-n^2\pi^2 \dfrac{4D_d t}{d_d^2} \right]$	Newman [23]
$\dfrac{c_0 - c_{fin}}{c_0 - c_{eq}} = 1 - \dfrac{3}{8} \sum\limits_{n=1}^{\infty} A_n^2 \exp\left[-\lambda_n \dfrac{64 D_d t}{d_d^2} \right]$	Kroning–Brink [24]
$A_1 = 1.32, A_2 = 0.73, \lambda_1 = 1.678, \lambda_2 = 9.83$	
$\dfrac{c_0 - c_{fin}}{c_0 - c_{eq}} = 1 - 2 \sum\limits_{n=1}^{\infty} B_n^2 \left[\dfrac{\lambda_n U t}{128(1 + \mu_d/\mu_c) d_d} \right]$	
$\lambda_1 = 2.866$	Handlos-Baron [25]

2.2). Based on experimental results, they concluded that for the oscillating droplets the best results were obtained with the model of Skelland and Wellek and the model of Rose and Kintner in Table 2.2. They are much better than the other relationships. The Handlos and Baron model was originally developed for describing mass transfer within the droplet. Later on it was extended to finite

Table 2.2. Comparison of models according to Brunson and Wellek [31]

$$Sh_d = \frac{k_d d_d}{D} = \frac{4}{\sqrt{\pi \tau_d}}; \quad \tau = \frac{4Dt_c}{d_d^2} \qquad \text{modified Higbie [18]}$$

with t_c droplet residence time in the column; Higbie modified by Brunson and Wellek [31]

$$Sh_d = \frac{2}{\sqrt{\pi}} \sqrt{\frac{2au}{D}} \quad \text{with } t_c = \frac{d_e}{U} \qquad \begin{array}{l}\text{Higbie modified by} \\ \text{Brunson and Wellek [31]}\end{array}$$

$$Sh_d = \frac{4}{\pi} \sqrt{\frac{d_d^2 \omega}{4D_d}} \quad \text{with } t = \frac{\pi}{\omega} \qquad \text{original Higbie [18]}$$

$$Sh_d = \frac{\lambda_1 Pe^{*2}}{768} \quad \text{with } \lambda_1 = 2.886, \quad Pe^* = \frac{d_e u_e}{D(1 + \mu_d/\mu_c)}$$

only for longer period of time τ Handlos-Baron [25]

$$Sh_d = 0.972 \frac{\lambda_1 Pe^*}{768} + \frac{0.3}{\tau}$$

also for shorter periods of time τ Olander [35]

$$Sh_d = 0.320 \, Re_c^{0.68} \, p^{0.10} \, \tau^{-0.14}$$

ethylacetate to water transfer   Skelland-Wellek [32]

$$Sh_d = 0.142 \, We^{0.77} \, p^{0.28} \, \tau^{-0.14}$$

water to ethylacetate transfer   Skelland-Wellek [32]

$$\text{where} \quad Re_c = \frac{d_d u_e}{\nu_c}, \quad P = \frac{\sigma^3 \rho_c^2}{g u_c^2 \Delta \rho}, \quad We = \frac{d_d u_e^2}{\sigma}$$

$$Sh_d = \frac{(k_d A_d)_d}{A_o D} = \frac{1}{\pi d_d t_c} \int_0^{t_c} \frac{A_d}{X} dt$$

t_c residence time of the droplet in the column   Rose–Kintner [33]
X interfacial resistance zone varies with time

$$Sh_d = \frac{4}{\pi} \sqrt{\frac{d_d^2 \omega}{4D}\left(1 + \varepsilon + \frac{3}{8}\varepsilon^2\right)}$$

with $\quad \varepsilon = \frac{A_{max}}{A_o} - 1$ Angelo et al. [34]

$$Sh_d = \frac{4}{\pi} \sqrt{\frac{d_d^2 \omega}{4D}(1 + h\varepsilon)} \qquad \text{Brunson-Wellek [31]}$$

h = 0.378 with t_c = const and h = 0.687 nach Beek-Kramers approach continuous phase Sherwood numbers Sh_c, either through the estimation of the eigenvalues [38, 39] or by the numerical solution of the governing differential equation [38, 40]. It was recently pointed out that with low continuous Sherwood numbers the Handlos-Baron model predicts mass transfer rates higher than those for a perfectly mixed droplet with an identical value of k_c [36]. This is due to the use of cylindrical geometry for a spherical droplet.

The Handlos-Baron model was extended to the high flux mass transfer regime by Korchinsky and Young [41]. At low solute concentrations and a low flux mass transfer regime, the Handlos-Baron model well predicts the performance of a rotating disc contactor. However, in a high flux regime, the mass transfer coefficients increase due to the rapidly decreasing interfacial tension. This effect is compensated by the increased axial dispersion in the continuous phase of the column.

A consecutive chemical reaction in solvent extraction can increase the capacity of the solvent phase, the extraction rate and the selectivity of the separation process. For this reason, mass transfer and chemical reaction can play a role in the extraction of metabolites or proteins too. Therefore, the interrelation between mass transfer and chemical reaction in single droplets, without taking the mass transfer resistance in the continuous phase into account, are also taken into consideration.

At the center of interest of the investigations is the acceleration (enhancement) of the mass transfer by consecutive chemical reaction. The enhancement factor is defined as the ratio of solute mole flux with chemical reaction to that without chemical reaction. For detailed treatment of this subject in gas/liquid systems, see the book of Danckwerts [44].

The Newman's mass transfer model [23] was extended by Crank [45] with irreversible reaction of the 1st order.

The partial differential equation (PDE) of the mass balance in a spherical droplet is given by Eq. (2.67):

$$\frac{\delta C}{\delta \tau} = \frac{\delta^2 C}{\delta R^2} + \frac{2}{R}\frac{\delta C}{\delta R} - k_R C = 0 \tag{2.67}$$

with the initial and boundary conditions: $C(R, 0) = 0$ and $C(0, \tau) = $ finite, where C is the solute concentration in the droplet, $R = r/a$ the dimensionless radial coordinate, r the radial coordinate, a the radius of the droplet, $\tau_d = D_d t/a^2$ the dimensionless time, $k_R = k_r a^2/D$ the dimensionless reaction rate constant, for the 1st order reaction, and $k_R = k_r a^2 C_{B0}/D$ the dimensionless reaction rate constant for 2nd order reaction.

If the reaction partner B in the droplet is in great excess, $C_B \approx C_{B0}$ is valid, thus C_B can be included into k_R; then we have a reaction apparently of the 1st order. In Table 2.3, the enhancement factor for the 1st order reaction in a droplet with laminar circulation is given.

Wellek, Andoe and Brunson [46] extended the Handlos/Baron model with a 1st order reaction by considering the circulation, but not the oscillation

Table 2.3. Enhancement factors for different systems

Laminar circulation in droplet with irreversible reaction 1st order:

$$\Phi = \frac{\sum_{n=1}^{\infty} U_n}{\sum_{n=1}^{\infty} \exp[-\tau(k_R + n^2\pi^2)]}$$ Crank [45]

$$U_n = \frac{k_R + n^2\pi^2 \exp[-\tau(k_R + n^2\pi^2)]}{k_R + n^2\pi^2}$$

Turbulent circulation in the droplet (without oscillation) with irreversible reaction 1st order:

$$\Phi = \frac{\sum_{n=1}^{\infty} A_n^2 \mu_n u_n}{\sum_{n=1}^{\infty} A_n \mu_n \exp[-\tau(k_R + 16\mu_n\tau)]}$$ Wellek et al. [46]

$$u_n = \frac{k_R + 16\mu_n \exp[-\tau(k_R + 16\mu_n)]}{k_R + 16\mu_n}$$

for eigen values A_n and μ_n ($n = 1 \ldots 7$) see Heertjes [47]

Turbulent (oscillating droplet with irreversible reaction of the 1st order:

$$\Phi = 1 + \frac{k_R}{\omega}\left[\frac{\pi/3 + 0.396\varepsilon}{1 + 0.687\varepsilon}\right]$$ Wellek et al. [46]

$$\varepsilon = \frac{A_{max}}{A_o} - 1$$

ω droplet oscillation frequency (rad /s)
A droplet oscillation amplitude

droplet phase. The PDE is given by

$$\frac{\delta C}{\delta \tau_d} = \frac{\delta^2 C}{\delta R^2} + \frac{2}{R}\frac{\delta C}{\delta R} + \frac{1}{R^2}\frac{\delta^2 C}{\delta \theta^2} + \frac{\cot\theta}{R}\frac{\delta C}{\delta \theta}$$
$$+ Pe_d\left[(1 - R^2)\cos\theta\frac{\delta C}{\delta R} + \frac{2R^2 - 1}{R}\sin\theta\right] - k_R C \qquad (2.68)$$

with the initial and boundary conditions:

$$C(R, \theta, 0) = 0$$
$$C(1, \theta, \tau_d) = 1$$
$$C(0, \theta, \tau_d) = \text{finite}$$

$$\frac{\delta C}{\delta \theta}(R, 0, \tau) = \frac{\delta C}{\delta \theta}(R, \pi, \tau_d) = 0$$

where θ droplet angular coordinate

$$Pe_d = \frac{1}{4}\left[\frac{\mu_c}{\mu_c + \mu_d}\right]\frac{d_d U}{D_d}$$

is the droplet Peclet-number, μ_c, μ_d the viscosities of continuous and dispersed phases and U the (relative) droplet velocity.

The enhancement factor for this case is given in Table 2.3. For the development of the enhancement factor in oscillating droplets, the model of Brunson and Wellek [31] was used, in which the droplet interfacial area function $A = A_0(1 + \sin^2 \omega')$ according to Angelo et al. [34] and the instantaneous mass transfer coefficient $k_d(t) = \sqrt{D/\pi t}(1 + k_r t)$ according to the penetration theory was applied [46].

The investigations with nonoscillating droplets indicate, if $k_R > 640$, the effect of the droplet Peclet-number (and hence the internal circulation) on the rate of mass transfer is negligible. The penetration theory solutions may be used for short contact times, and the stagnant sphere series solutions may be used for longer contact times.

When $k_R < 640$, in oscillation the droplet effect of droplet Peclet-number on the rate of mass transfer is significant. However, if $\tau_d < 10^{-3}$, changes in the Pe-number have virtually no effect on the rate of mass transfer into the droplet and the penetration theory relationships may be employed. In rapidly oscillating droplets, the effect of chemical reaction may be only of secondary importance to he effect of surface renewal [46].

Mass transfer with instantaneous second order reaction in drops with laminar circulation was experimentally investigated by Tyroler et al [48] and Halwachs [49] by means of the reaction of acetic acid in the continuous cyclohexanol phase and NaOH in the aqueous droplet phase with phenolphthalein as indicator. The investigations were carried out with circulating and stagnant drops. With instantaneous reaction a reaction front is formed [44], at which the violet color of phenolphthalein disappears. With a sequence of photographs [48] and with moving pictures [49] the variation of the reaction front was observed. Tyroler et al. [48] evaluated these pictures quantitatively. A comparison with the theory was carried out by Brunson and Wellek [50]. Their model is not dealt with here because of its complexity. Only some results and conclusions are taken into account. A reaction with the following stoichometric equation

$A + zB \rightarrow$ Products

has been included. The total mol mass of solute A (acetic acid) transferred into the drop divided by the product of the droplet volume and the surface concentration of component A was calculated. In Fig. 2.10, this value is plotted as a function of the dimensionless time τ_d at different k_R values. With increasing k_R the transferred solute mass increases. At diffusivity ratio of A to B: $R_D = 1$, the ratio of concentration of A at the surface to the initial concentration of B in the droplet: $R_C = 0.2$ and Pe = 40, the total mass of A transferred into the drop attains for a large τ_d the value of 6, which is equal to

$1 + 1/R_C$.

The diffusivity ratio R_D has only a slight effect on the mass transferred at a low reaction number. With a high reaction number ($k_R = 640$) its effect is

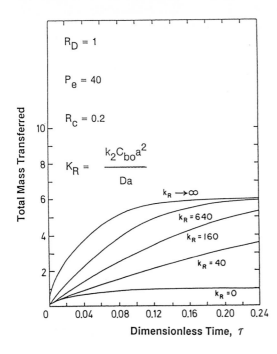

Fig. 2.10. Total mass transferred as a function of the dimensionless time τ_d at different reaction numbers k_R [50]

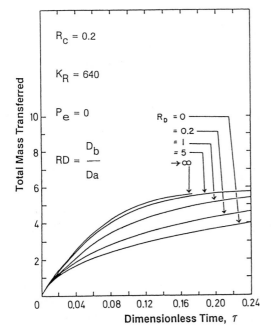

Fig. 2.11. Total mass transferred as a function of the dimensionless time τ_d at different diffusivity ratios $R_D = D_B/D_A$ [50]

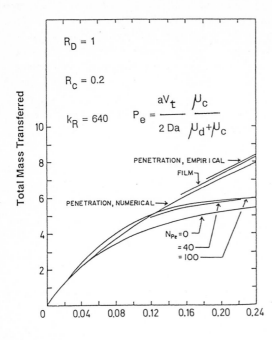

Fig. 2.12. Total mass transferred as a function of the dimensionless time τ_d at different droplet peclet numbers Pe_d [50]

considerable (Fig. 2.11). The effect of the diffusivity ratio decreases as Pe increases.

Figure 2.12 shows the effect of the internal circulation on the total mass transferred at $R_D = 1$, $R_C = 0.2$ and $k_R = 640$ predicted by the model. For $\tau_d < 10^{-3}$, the difference in the total mass transferred – calculated by the model of Brunson and Wellek – differs from the one predicted by either the film or penetration model by less than 5%. This is the range of practical importance. At higher contact times, the assumption of the flat interface, which is basic to both, the film and the penetration model, is no longer valid. At $\tau_d > 0.02$, both film and penetration model predict values of total mass transfer greater than possible due to neglecting the depletion of reactant B (NaOH) within the droplet [50].

2.3.2 Mass Transfer in the Continuous Phase

The models used for the calculation of mass transfer in the continuous phase are partly identical to those that are developed for gas/liquid reactions. A comprehensive review of the latter is given by Danckwerts [44]. The reason for this coincidence is that they partly disregard the specific bubble/droplet fluid dynamics and the curvature of the interface, which allows considerable reduction of the PDEs and the evaluation of the analytical solutions.

A critical boundary condition is the solute concentration at the interface, which is usually assumed to be constant. This yields quasi steady-state solutions for which the variations in the bulk are more rapid than those of the boundary conditions. This assumption could hold true for the pure mass transfer process. For chemical reaction coupled mass transfer processes this assumption is often not true. Quasi steady-state solutions with different boundary conditions can be used for the description of the transfer processes.

For gas/liquid reactions, five regimes can be distinguished [51], if one disregards the transient regimes between them. These regimes can also exist in liquid/liquid systems, if the mass transfer resistance in one of the phases is negligible. Numerous relations are known for the rigid drop, which have the form

$$Sh_c = \text{const. } Re_c^{0.5} Sc_c^{0.33} \tag{2.68a}$$

or

$$Sh_c = 2 + \text{const. } Re_c^{0.5} Sc_c^{0.33} \tag{2.68b}$$

According to Linton and Sutherland [59] and Thorsen [60] the constant in Eq. (2.68a) equals 0.582. Keey and Glen [63] recommend the value 0.76. for the constant in Eq. (2.68b).

For nonrigid drops, Eq. (2.68a) was also recommended, however, with a different numerical value of the constant. Boussinesq [62] came to the evaluation of the exponent 0.5 of Sc and to the constant 1.13, Garner and Tayeban [43] recommend 0.6.

For high Reynolds numbers, the mass transfer resistance external to a spherical drop was evaluated numerically by Weber [64] by using the thin concentration boundary-layer assumption and the interfacial velocity given by Harper and Moore [65]. His results can be put into the following form

$$Sh = \frac{2}{\sqrt{\pi}} \left[1 - \frac{1}{Re_c^{0.5}} (2.89 * B) \right]^{0.5} Pe_c^{0.5}, \tag{2.69}$$

where the Sherwood-, Reynolds and Peclet numbers are based on the properties of the external phase:

$$Sh_c = \frac{k_c d_d}{D_c}, \quad Re_c = \frac{d_d U}{v_c}, \quad Pe_c = \frac{d_d U}{D_c},$$

where k_c is the mass transfer coefficient in the continuous phase and the constant B is a function of $x = \mu_d/\mu_c$ and $y = \delta_d/\delta_c$, For $x \leq 2$ and $0 < y \leq 4$, the following approximate form applies:

$$Sh_c = \frac{2}{\sqrt{\pi}} [1 - Re_c^{-0.5}(2.89 + 2.15 x^{0.64})]^{0.5} \tag{2.70}$$

with an error less than 5% in $Sh_c/Pe_c^{0.5}$. Heertjes and De Nie [61] gave an excellent review of these equations.

Huang and Kuo [52] examined the nonsteady state extraction coupled with the first order reaction based on the film-penetration model of Toor and

Marcello [21]. The mass balance of solute A is given by the PDE:

$$\frac{\delta C_A}{\delta t} = D_A \frac{\delta^2 C_A}{\delta x^2} - k_r C_A \quad \text{for } 0 \leq x \leq L, \tag{2.71}$$

with the initial and boundary conditions:

$$x = 0 \qquad t > 0 \qquad C_A = C_{Ai}$$

$$x = L \qquad t > 0 \qquad C_A = C_{AL}$$

$$x > L \qquad t = 0 \qquad C_A = C_{AL}.$$

The molar flux of A is given by Eq. (2.72):

$$j_A = \frac{\sqrt{D_A}}{\pi t} [C_{Ai} \exp(-k_r t) + C_{AL}(\exp(-1/4\alpha)$$

$$+ 2\exp[-(1/4\alpha + k_r t) - 1)]$$

$$+ C_{Ai} \sqrt{k_r D_A} \, \text{erf} \sqrt{k_r t}$$

$$+ C_{Ai} \sum_{n=0}^{\infty} \left[2 \frac{\sqrt{D_A}}{\pi t} \exp - [(n^2/\alpha k_r t + k_r t)] \right.$$

$$+ \sqrt{k_r D_A} \left\{ \exp(-2n/\sqrt{\alpha}) \, \text{erfc} \left(\frac{n}{\sqrt{\alpha k_r t}} - \sqrt{k_r t} \right) \right.$$

$$\left. - \exp(-2n/\sqrt{\alpha}) \, \text{erfc} \left(\frac{n}{\sqrt{\alpha k_r t}} + \sqrt{k_r t} \right) \right\} \right]$$

$$+ C_{AL} \sum_{n=1}^{\infty} \frac{\sqrt{D_A}}{\pi t} \left\{ (-1)^n \exp \left[-\frac{(n+1)^2}{4\alpha} \right] \right.$$

$$+ 2\exp \left[-\frac{(2n+1)^2}{4\alpha} + k_r t \right] - (1)^n \exp \left[-\frac{n^2}{4\alpha} \right] \right\}$$

$$+ \sqrt{k_r D_A} \left\{ \exp \left[-\frac{2n+1}{\sqrt{\alpha}} \right] + \text{erfc} \left[\frac{2n+1}{2\sqrt{\alpha k_r t}} + \sqrt{k_r t} \right] \right.$$

$$\left. - \exp \left[-\frac{2n+1}{\sqrt{\alpha}} \right] \text{erfc} \left[\frac{2n+1}{2\sqrt{\alpha k_r t}} - \sqrt{k_r t} \right] \right\} \right] \tag{2.72}$$

where $\alpha = D_A/k_r L^2 = M^{-1}$, L = diffusion distance, x = distance from the interface.

The disadvantage of Eq. (2.72) is that C_{Ai} is unknown in the steady state process.

If $L \to \infty$ and $C_{Ai} \to 0$, Eq. (2.72) is reduced to:

$$j_A = C_{Ai} \left[\frac{\sqrt{D_A}}{\pi t} \exp(-k_r t) + \sqrt{k_r D_A} \, \text{erf} \sqrt{k_r t} \right], \tag{2.73}$$

which holds true for the penetration model. Furthermore, with $k_r \to 0$, the Higbie model

$$j_A = \frac{\sqrt{D_A}}{\pi t}(C_{Ai} - C_{AL})$$
(2.74)

is obtained. The mean flux can be calculated from Eq. (2.72):

$$j_A = \sqrt{k_r D_A (1 + \beta)}\left[\left(C_{A1} - \frac{\beta C_{AL}}{1 + \beta}\right)\coth\frac{\sqrt{1 + \beta}}{\alpha}\right.$$

$$\left. - \frac{C_{AL}}{1 + \beta}\operatorname{cosech}\frac{\sqrt{1 + \beta}}{\alpha}\right]$$
(2.75)

where $\beta = s/k_r$, s surface renewal frequency.

If α disappears, Eq. (2.75) yields the Danckwerts' surface renewal model:

$$j_A = \sqrt{D_A(k_r + s)}\left(C_{Ai} - \frac{sC_{AL}}{k_r + s}\right)$$
(2.76)

If β disappears in Eq. (2.75), Eq. (2.77) of Hatta results:

$$j_A = \frac{D_A}{L}\frac{1}{\sqrt{\alpha}}\left[C_{Ai}\coth\frac{1}{\sqrt{\alpha}} - C_{AL}\operatorname{cosech}\frac{1}{\sqrt{\alpha}}\right]$$
(2.77)

If $k_r = 0$, Eq. (2.75) reduces to the film penetration model of Toor and Marcello:

$$j_A = \sqrt{D_A s}\left[\frac{2}{1 + \exp(-2\sqrt{sL^2/D_A})} - 1\right](C_{A1} - C_{AL})$$
(2.78)

By means of this film penetration model, the following enhancement factor can be calculated:

$$\Phi = \sqrt{1 + M\left(\coth\frac{1}{\sqrt{\gamma}}\right)^2}\frac{\tanh(1/\gamma)^{0.5}}{\coth\left[(1/\gamma)(1 + M\{\coth(1/\gamma)^2\}125\right]^{0.5}}$$
(2.79)

where $\gamma = \alpha/\beta = D_A/sL^2$.

With $\gamma \to 0$

$$\Phi = (1 + M)^{0.5}$$
(2.80)

is obtained, which holds true for the surface renewal model.

With $\gamma \to \infty$

$$\Phi = M^{0.5}\coth M^{0.5}$$
(2.81)

is obtained, which holds true for the film model. Rapid chemical reactions influence the concentration gradients in the diffusion layer as significant conversion is attained in it. In the case of irreversible first order reaction in regime 3 (Table 2.4) the enhancement factor is given by Eq. (2.81) again [49].

Table 2.4. Five regimes of gas/liquid reactions [44, 51]

Range	Influence of					
	rel. vol. of liqu. $a + \phi(1)$	spec. interf. $a(cm^{-1})$	phys. abs. coeff $k_p(cm\,s^{-1})$	react. rate $(R)^S$	enhancement $\Phi(1)$	absorb. rate $(mol\,cm^{-3}\,s^{-1})$
1	+	−	−	$s = 1$	1	$a\phi$
2	−	+	+	$s = 0$	1	$ak_p(c_{Ai} - c_{Aeq})$
3	−	+	−	$s = 0.5$	$\sqrt{\alpha}$	$ak_c(c_{Ai} - c_{Aeq})$
4	−	+	+	$s = 0$	$1 + \dfrac{D_B c_B^o}{D_A c_{Ai} q}$	$\dfrac{ak_p c_B^o}{q}$
5	−	+	−	$s = 0$		$ak_g c_{Ag}$

Explanations to Table 2.4: + proportional, − independent B is present in the liquid phase with the bulk concentration c_B^o, ϕ liquid volume/gas-liquid interfacial area (cm), a specific gas-liquid interfacial area (cm^{-1}), physical absorption rate of A: $Q_{Ap} = a\,k_p(c_{Ai} - c_A^o)$, chemical absorption rate of A: $Q_{Ac} = a\,k_c(c_{Ai} - c_{Aeq})$, c_{Ai}^o, c_A^o, c_{Aeq} the concentrations of solute A transferred from the gas phase into the liquid phase at the gas liquid interface, in the bulk liquid without chemical reaction and in the liquid bulk with chemical reaction in equilibrium, c_{Ag} concentration of A in the bulk gas phase, k_g mass transfer coefficient through the gas film, q stoichiometric coefficient, D_A, D_B diffusivities of A and B in the liquid phase, α is the ratio of the equivalent diffusion time $t_D = D_A/(k_p)^2$ to the equivalent reaction time $t_R = (c_{Ai} - c_{Aeq})/R$.

The extraction enhancement starts at about $M^{0.5} \approx 0.5$ and for $M^{0.5} \geq 2$ the relationship between Φ and $M^{0.5}$ is linear.

2.3.3 Resistances of Both Phases under Consideration

Generally, the mass transfer resistances in both phases have to be taken into account. Which resistance dominates depends on the ratio of the diffusivities of the solute in the two phases $R_D = D_d/D_C$ and the partition coefficient $P_C = (C_{Ad}/C_{Ac})_{equil}$. The resistance lies mainly in the dispersed phase d if

$$R_D \ll 1 \quad \text{and} \quad P_C < 1,$$

and mainly in the continuous phase c if

$$R_D \gg 1 \quad \text{and} \quad P_C > 1,$$

and in both phases if

$$R_D \approx 1 \quad \text{and} \quad P_C \approx 1.$$

According to Brauer [66] the criterion $P_c R_D^{0.5}$ is decisive for the dominance of the resistances. Only for $0.03 < P_c R_D^{0.5} < 30$ should both resistances be considered.

For $P_c R_D^{0.5} < 0.03$, only the resistance in the droplet phase has to be dealt with, and for $P_c R_D^{0.5} > 30$, only the resistance in the continuous phase must be taken into account.

Brounshtein and Fishbein [67] have shown that the mass-transfer process between the drop and the continuous phase is in a nonsteady state, when the dispersed phase resistance controls. Therefore, the non-steady state differential equation of mass transfer of the solute A has to be applied [64, 68–70]. The mass transfer balance of A in the droplet phase is given by

$$\frac{\delta C_{Ad}}{\delta t} + u_{rd}\frac{\delta C_{Ad}}{\delta r} + \frac{u_{\theta d}}{r}\frac{\delta C_{Ad}}{\delta \theta}$$

$$= D_d\left[\frac{\delta^2 C_{Ad}}{\delta r^2} + \frac{2}{r}\frac{\delta C_{Ad}}{\delta r} + \frac{1}{r^2\sin\theta}\frac{\delta}{\delta\theta}\left(\sin\theta\frac{\delta C_{Ad}}{\delta\theta}\right)\right] \qquad (2.82a)$$

Here C_{Ad} is the concentration of the solute A in the droplet phase, r is the radial coordinate, θ is the angular coordinate measured counterclockwise from the upstream stagnation point; u_{rd} is the radial velocity along r, and $u_{\theta d}$ is the tangential velocity along θ.

The mass balance of A in the continuous phase is:

$$\frac{\delta C_{Ac}}{\delta t} + u_{rc}\frac{\delta C_{Ac}}{\delta r} + \frac{u_{\theta c}}{r}\frac{\delta C_{Ac}}{\delta \theta}$$

$$= D_c\left[\frac{\delta^2 C_{Ac}}{\delta r^2} + \frac{2}{r}\frac{\delta C_{Ac}}{\delta r} + \frac{1}{r^2\sin\theta}\frac{\delta}{\delta\theta}\left(\sin\theta\frac{\delta C_{Ac}}{\delta\theta}\right)\right] \qquad (2.82b)$$

Here C_{Ac} is the concentration of the solute A and u_{rc} and $u_{\theta c}$ are the corresponding velocities in the continuous phase. The initial conditions are for $t = 0$:

$$C_{Ad} = C_{Ad\theta} \quad \text{for each } r$$

$$C_{Ac} = C_{Ac\infty}$$

and the boundary conditions:

for $\theta = 0$,

$$\delta C_{Ad}/\delta\theta = 0 \quad \text{and} \quad \delta C_{Ac}/\delta\theta = 0,$$

for $r = 0$,

$$\theta = \pi/2$$

$$\delta C_{Ad}/\delta r = 0,$$

for $r = \infty$

$$C_{Ac} = C_{Ac\infty} = \text{constant.}$$

Equilibrium prevails at the interface:

at $r = a$

$$(C_{Ad})_{r=a} = P_c(C_{Ac})_{r=a}$$

and the solute is transferred across the interface by pure diffusion:

at $r = a$

$$D_d(\delta C_{Ad}/\delta r)_{r=a} = D_c(\delta C_{Ac}/\delta r)_{r=a}.$$

An analytical solution of Eq. (2.82) is possible, if one takes into account short contact times only. In that case, the distribution of the solute concentration in the droplet does not depend on the flow, and convective terms in Eq. (2.82) disappear.

The solution for this case is given by Brauer [66]:

$$\xi_d = 1 - \frac{6}{\sqrt{\pi}} \frac{1}{1 + P_c R_d^{0.5}} \tau_d^{0.5} \qquad\qquad (2.82c)$$

where

$$\xi_d = \frac{c_{Ad} - P_c C_{Ac}}{C_{Ad0} - P_c C_{Ac}}$$

is the dimensionless concentration and $\tau_d = tD_d/a^2$ the dimensionless time (droplet Fourier number). An analytical solution for the general case is not known. The numerical solution of Eq. (2.82) is given by Brauer [66] which is restricted to $0 < Re < 1$. For $Re > 1$ it is assumed that the mass transfer is a quasi-steady state process, and that the mass transfer coefficient in the continuous phase is constant and can be calculated by the equations of steady-state mass transfer. By this assumption it is possible to decouple the differential equations of the two phases. In that case only the solution of the balance equation in the droplet phase is needed. The constant resistance in the continuous phase appears only as a boundary condition. This resistance is defined by means of a Biot-number:

$$Bi = k_c 2a/D_c,$$

where k_c is the fractional mass transfer coefficient in the continuous phase.

If the resistance in the droplet phase dominates, the Bi-number approaches infinity. In the other limiting case with dominating resistance in the continuous phase $Bi \to 0$. For systems at rest ($Re \approx 0$), the quasi steady-state treatment of Gröber [71, 72] for nonsteady state heat transfer can also be applied for non steady state mass transfer [69]:

$$\xi_d = 6 \sum_{n=1}^{\infty} B_n \exp(-\lambda_n^2 \tau_d). \qquad\qquad (2.83)$$

The eigenvalues B_n and λ_n are given in [71].

2.3.4 Comparison of the Models

To compare the models, they are brought into the same form. If the resistance in the continuous phase can be neglected, Eq. 2.83 of Brauer is reduced to

$$\xi_d = 1 - \frac{6}{\sqrt{\pi}} \pi_d^{0.5}. \tag{2.84}$$

Therefore, the mean concentration depends only on the dimensionless time τ_d in Eq. (2.84).

For drops at rest, the mass transfer inside the drop occurs only by radial diffusion [23]:

$$\xi_d = \frac{6}{\sqrt{\pi}} \sum_{n=1}^{\infty} \frac{1}{n^2} \exp(-n^2 \tau_d). \tag{2.85}$$

In circulating drops, the model of Kronig and Brink can be applied [24]:

$$\xi_d = \frac{3}{8} \sum_{n=1}^{\infty} B_n^2 \exp(-\lambda_n 16 \tau_d). \tag{2.86}$$

The numerical values of λ_n and B_n are given by Heertjes et al. [47].

For well mixed droplets, the model of Handlos and Baron can be employed [25]:

$$\xi_d = \sum_{n=1}^{\infty} 2B_n^2 \exp\left(-\lambda_n \frac{1}{64(1+x)} \tau_d Pe_d\right). \tag{2.87}$$

Handlos and Baron took into account only the first term of the series with $2B_1^2 = 1$ and $\lambda_1 = 2.88$. The latter was corrected to 2.866 by Skelland and Wellek [73]. Olander [35] pointed out that this assumption is only true for long contact times. For short contact times he proposed the approximation with $2B_n = 0.64$ and $\lambda_1 = 2.80$.

In Fig. 2.13, the mean concentrations ξ_d are plotted as a function of τ_d according to the model of Brauer (analytical solution of Eq. (2.84)), Brauer (numerical solution of Eq. (2.82)), solutions of Newman (Eq. (2.85)), Kronig and Brink (Eq. (2.86)) and Handlos-Baron (Eq. (2.87)) for $2a = 0.31$ cm, $U = 7$ cm s^{-1}, and $D_d = 2.19 \cdot 10^{-5}$ cm^2 s^{-1}.

A comparison of the numerical results of Brauer with the results calculated by Eq. (2.85) indicates that it is necessary to consider 20 terms in this equation to obtain satisfactory results [68].

The influence of the resistance in the continuous phase on the extraction process is shown in Fig. 2.14 based on the model of Gröber [72] (Eq. 2.84). One can recognize that with increasing mass transfer resistance in the continuous phase (decreasing Bi number) the rate of extraction process is considerably reduced.

In Fig. 2.15, the same effect is shown by means of the model of Kronig and Brink, which was extended to the continuous phase by Elzinga and Banchero [75]. One can observe that for the small dimensionless time τ_d and Biot numbers the calculated concentration in the drop is too large. This can be attributed to the nonadequate approximation of only four terms in Eq. (2.86). This equation can only be used for large τ_d values.

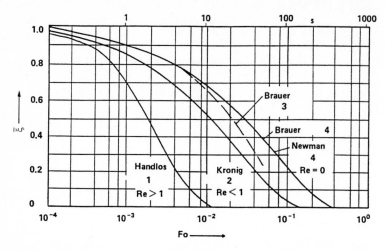

Fig. 2.13. Dimensionless concentration of solute ξ as function of the dimensionless time τ_d for different Reynolds numbers with mass transfer resistance only in the droplet phase (1) calculated by model of Handlos Baron [25], (2) calculated by the model of Kronig and Brink [24], (3) calculated by the model of Brauer [68], (4) calculated by Brauer (numerical) [68] and by the model of Newman [23] with 20 terms [74]

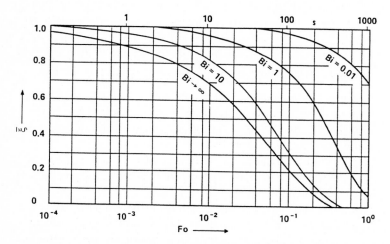

Fig. 2.14. Dimensionless concentration ξ as a function of the dimensionless time τ_d with mass-transfer resistance in both the phases. (Calculated according to Gröber [71, 72] [74]

Skelland and Wellek [73] extended the Handlos-Baron model to systems with comparable resistances in the two phases. They developed the eigenvalues of the Handlos-Baron model Eq. (2.87) as a function of the resistance number h:

$$h = 512 \, (1 + x) \, (k_c / U P_c). \tag{2.88}$$

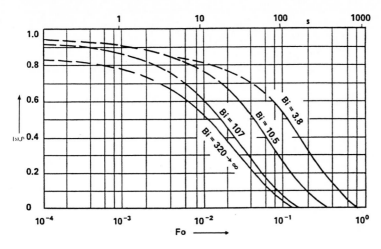

Fig. 2.15. Dimensionless concentration ξ as a function of the dimensionless time τ_d with mass transfer resistance in both phases for $Re_d < 1$. (Calculated according to Elzinga and Banchero [75] [74]

They assumed that only λ_n is a function of h. By using the λ_n eigenvalues published by Skelland and Wellek [73] the ξ_d values deviate for a short period of time from ones published by Patel and Wellek two years later [40]. The value of ξ_d for h $\rightarrow \infty$ agrees with the approximation of Olander [35] for short contact times. Figure 2.16 shows some results according to Patel and Wellek for Re > 1 [40]. Again, the strong dependence of the extraction process on the mass

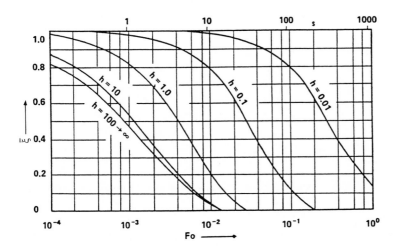

Fig. 2.16. Dimensionless concentration ξ as a function of the dimensionless time τ_d with mass transfer resistance in both phases for $Re_d < 1$ (Calculated according to Patel and Wellek [40] [74]

transfer resistance in the continuous phase is obvious. In Table 2.5 the validity ranges of the different models are shown.

Of the extraction processes coupled with chemical reactions occuring in systems in which the resistances of both phases have to be taken into account only the instantaneous irreversible bimolecular reaction (regime 4 in Table 2.4) is taken into consideration. The enhancement factor of these systems is given by [44]:

$$\Phi = 1 + (D_B C_B / D_A C_A) \tag{2.89a}$$

or

$$\Phi = 1 + (D_C C_C / D_d C_d), \tag{2.89b}$$

where solute A transferred from the drop (d) to the continuous (c) phase, in which the reaction partner B is dissolved. The reaction front is in the diffusion layer of the continuous phase (Fig. 2.17) and it is shifted towards the interface with increasing reaction rate. If the latter is very high, the reaction front is at the interface, thus the mass transfer resistance of solute A in the continuous phase approaches zero.

If solute A is originally in the droplet phase A, the reactant B is in the continuous phase B and the reaction occurs in the diffusion layers of phases A and B (Fig. 2.18), the mass balance of components A and B in the phase B is given by:

$$D_{AB} \frac{dC_{AB}^2}{dx^2} = z_{AB} k_{rB} C_{AB}^{pB} C_{BB}^{qB} \tag{2.90a}$$

$$D_{BB} \frac{dC_{BB}^2}{dx^2} = z_{BB} k_{rB} C_{AB}^{pB} C_{BB}^{qB} \tag{2.90b}$$

Table 2.5. Summary of the models for mass transfer between the two phases without chemical reaction [74]

	Mass transfer resistance in continuous phase	Mass trasfer resistance in dropet phase	Mass transfer resistances in both of the phases	
			Numerical solutions	Quasistationary
Re = 0		Newman [23] Brauer [68]	Brauer [68]	Gröber [71]
Re < 1	Levich [26] Heertjes (review) [61]	Kronig and Brink [24] Schmidt-Traub cited in [74]	Schmidt-Traub cited in [74]	Elzinga and Banchero [75]
Re > 1	Heertjes (review) [61] Weber [64]	Handlos and Baron [25]		Skelland and Wellek; cited in [69] Patel and Wellek [40]

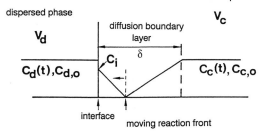

2.17. Schematic description of the concentration profiles for a bimolecular irreversible instantaneous reaction in the diffusion boundary layer [49]

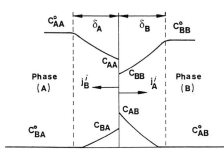

2.18. Schematic description of the concentration profiles for a bimolecular irreversible reaction in both phases [49]

with the boundary conditions:

$$C_{AB} = C_{AB}^i \quad \text{and} \quad dC_{BB}/dx = J_B/D_{BB} \quad \text{at } x = 0 \tag{2.91a}$$

$$C_{AB} = 0 \quad \text{and} \quad C_{BB} = C_{BB}^o \quad \text{at } x = \delta_B. \tag{2.91b}$$

For the enhancement factor in phase B

$$\Phi_{AB} = \sqrt{M_B P_B^{qB}} \coth(\sqrt{M_B P_B^{qB}}) \tag{2.92}$$

is obtained, if the reaction coupled mass transfer equations are linearized [53]. In Eq. (2.92) $P_B = C_{BB}^i / C_{BB}^o$ and

$$M_B = \frac{2}{b_B + 1} k_{rB} (C_{AB})^{pB-1} (C_{BB}^o)^{qB} \frac{D_{AB}}{k_{AB}^2}. \tag{2.93}$$

For the enhancement factor in phase A, a similar relationship holds true [53].

If components B and C are insoluble in the organic solvent phase and the component A and the product ABC in the aqueous phase, the reaction can only occur at the interface, and the product is enriched in the organic phase. If the process is an instantaneous reaction (ion reaction), the extraction rate is controlled by the mass transfer of the components from the bulk to the interface. This system was taken into consideration by Reschke et al. [54]:

$$A_o + B_a + C_a \rightleftharpoons ABC_o. \tag{2.94}$$

The quasi steady-state solution of the rate equation is given for the solute B as follows:

$$-\frac{dC_B}{dt} = k_B a\left[C_B + \frac{1}{2}\left(\frac{k_A}{k_{ABC}KC_C} + \frac{k_A C_A}{k_B} - C_A\right)\right.$$
$$\left. - \sqrt{\frac{1}{2}\left(\frac{k_A}{k_{ABC}KC_c} + \frac{k_A C_A}{k_B} - C_B\right)^2 + \frac{k_A C_A}{k_{ABC}KC_c} + \frac{k_A C_{ABC}}{k_B KC_c}}\right].$$

(2.95)

This relationship holds true for the ion pair extraction of organic ions in water with organic carriers insoluble in the aqueous phase. Here, k_A, k_B, k_C, and k_{ABC} are the mass transfer coefficients of components A and ABC in the organic and B and C in the aqueous phases. C_{Ao}, C_{ABCo} are the concentrations of A and ABC in the organic phase and C_{Ba} and C_{Ca} the concentrations of B and C in the aqueous phase.

$$K = \frac{C_{ABCoi}}{C_{Aoi} \cdot C_{Bai} \cdot C_{Cai}},$$

the equilibrium constant.

In a similar case, A and ABC are insoluble in the aqueous phase, B and C are insoluble in the organic phase, but with slow rate determining reaction, an interfacial adsorption model can be used, which was developed by Völkel et al. [55]. However, this model has only been used for copper extraction with a carrier in a liquid membrane and no application of this reaction type is known in biotechnology up to now; therefore, it is not under consideration here.

2.3.5 Symbols for Sect. 2.3

A	droplet interfacial area	L^2
A_o	mean value of A	L^2
a	radius of the droplet	L
Bi	$= k_c 2a/D_c$ Biot number of the continuous phase	1
C	solute concentration in the droplet	$M\,L^{-3}$
C_A, C_B, C_{ABC}	concentrations of A, B and ABC, (Eq. (2.95)	$M\,L^{-3}$
C_A	molar concentration of A	$M\,L^{-3}$
C_{Ai}	C_A at $x = 0$	$M\,L^{-3}$
C_{AL}	C_A at $x = L$	$M\,L^{-3}$
C_{Ad}	concentration of A in the droplet phase	$M\,L^{-3}$
C_{Ac}	concentration of A in the continuous phase	$M\,L^{-3}$
C_{AB}	concentration of A in phase B	$M\,L^{-3}$
C_B	concentration of the reaction partner B in the droplet	$M\,L^{-3}$
C_{Bo}	C_B at $t = 0$	$M\,L^{-3}$

C_{BB}	concentration of B in phase B	$M L^{-3}$
D_{AB}	diffusivity of component A in phase B	$L^2 T^{-1}$
D_{BB}	diffusivity of component B in phase B	$L^2 T^{-1}$
D_c	solute diffusivity in the continuous phase	$L^2 T^{-1}$
D_d	solute diffusivity in the droplet phase	$L^2 T^{-1}$
d_d	droplet diameter	L
g	acceleration of gravity	$L T^{-2}$
h	resistance number Eq. (2.88)	1
j_A	molar flux of A	$M L^{-2} T^{-1}$
$k_A, k_B, k_C,$		
k_{AB}, k_{ABC}	mass transfer coefficients of A, B, C, AB and ABC	$L T^{-1}$
k_c	mass transfer coefficients in the continuous phase	$L T^{-1}$
$k_d(t)$	instantaneous mass transfer coefficient	$L T^{-1}$
k_R	$= k_r a^2/D$ dimensionless reaction rate constant for the 1st order reaction	1
k_R	$= k_r a^2 C_{Bo}/D$ dimensionless reaction rate constant for the 2nd order reaction	1
k_r	reaction rate constant for 1st order reaction	T^{-1}
k_r	reaction rate constant for 2nd order reaction	$L^3 M^{-1} T^{-1}$
k_{rB}	reaction rate constant for 1st order reaction in phase B	T^{-1}
L	diffusion distance	L
	(diffusion film thickness)	L
M	$= \alpha^{-1}$	1
Pe_c	$= d_d U/D_c$ Peclet number of the continuous phase	1
Pe_d	$= 1/4 \left[\dfrac{\mu_c}{\mu_c + \mu_d} \right] \dfrac{d_d U}{D_d}$ droplet Peclet number	1
R	$= r/a$ dimensionless radial coordinate	1
R_D	$= D_d/D_c$ ratio of diffusion coefficients in the droplet and continuous phases	1
Re_c	$= d_d U/v_c$ Reynolds number of continuous phase	1
r	radial coordinate	L
Sc_c	$= v_c/D_c$ Schmidt number of continuous phase	1
Sh_c	$= k_c d_d/D_c$ Sherwood number of continuous phase	1
s	surface renewal frequency	T^{-1}
U	relative droplet velocity	$L T^{-1}$
U_o	relative velocity of rigid droplets according to Stokes	$L T^{-1}$
U_{rc}	radial velocity in the continuous phase along r	$L T^{-1}$
$U_{\theta c}$	tangential velocity in the continuous phase along θ	$L T^{-1}$
u_{rd}	radial velocity in the droplet along r	$L T^{-1}$
$u_{\theta d}$	tangential velocity in the droplet along θ	$L T^{-1}$
x	distance from the interface	1
x	$= \mu_d/\mu_c$ viscosity ratio	1
y	$= \rho_d/\rho_c$ density ratio	1

α	$= D_A/k_r L^2 = M^{-1}$	1
β	$= s/k_r$	1
δ_B	film thickness in phase B	L
$\lambda_1, \lambda_2, \lambda_3$	retardation coefficients Eq. (2.65)	1
θ	droplet angular coordinate measured counterclockwise from the upstream stagnation point	1
μ_c	dynamic viscosity of the continuous phase	$M L^{-1} T^{-1}$
μ_d	dynamic viscosity of the droplet phase	$M L^{-1} T^{-1}$
v_c	kinematic viscosity of the continuous phase	$L^2 L^{-1}$
v_d	kinematic viscosity of the droplet phase	$L^2 L^{-1}$
ξ_d	dimensionless concentration Eq. (2.82b)	1
ρ_c	density of the continuous phase	$M L^{-3}$
ρ_d	density of the droplet phase	$M L^{-3}$
τ	$= Dt/a^2$ dimensionless time	1
τ_d	$= D_d t/a^2$ dimensionless time (droplet Fourier number)	1
Φ	enhancement factor	1
Φ_{AB}	enhancement factor in phase B	1
ω	cyclical frequency	T^{-1}

2.4 Experimental Techniques

2.4.1 Holdup and Drop Size Distribution

In analog to gas-liquid systems the holdup can be measured off-line after separation of the phases by determining the phase volumes or on-line during mass transfer by the absorption of X-rays or γ-rays [76, 77]. The local relative holdup can also be measured by optical, electro-optical probes or in the salt-containing aqueous phase with electrical conductivity probes. These techniques were developed for the determination of the bubble or droplet size:

a) the electrical conductivity probe utilizes the difference in conductivity between the aqueous and organic phases [78, 79],
b) when using an electro-optical probe, the two phase flow is sucked out through a capillary tube, in which the volume of the dispersed phase is measured by two photelectric gates [78, 80],
c) the glass fiber probe utilizes the differences in the refractive indexes of the aqueous and organic phases [81].

Usually, compound probes are used which consist of an array of two to five probes to improve the quality of the measurements. However, the flash photography of the droplets and the semiautomatic evaluation of the photographs is the most popular technique for the determination of the droplet-size distribution [78].

2.4.2 Mass Transfer Coefficient of Single Drops

The two most popular methods are very simple: After the formation of a drop on the tip of a capillary, holding it at the tip for mass transfer and its withdrawal, the solute concentation is determined in the droplet phase. The integral mass transfer coefficient can be calculated from the difference of the initial and final concentrations of the solute and the droplet surface area [82–84]. However, the droplet formation and withdrawal strongly influence the results. Furthermore, the hydrodynamics of the droplet held on the tip of the capillary differs considerably from that which moves freely. Therefore, a second method was developed which lets the droplet rise or fall in the continuous phase at rest.

Single droplets are consecutively formed by a capillary. After the detachment of the single droplets from the tip of the capillary, they rise (or fall) in the continuous phase. At the top (or at the bottom) of the column the droplet phase forms a layer in which the droplets are collected and analysed [30, 84–86].

In order to eliminate the end effects (mass transfer during droplet formation and collection), the measurements are carried out using different distances between the capillary tip and the interface of the phases. If the continuous phase is at rest (no convection prevails), the solute concentration is uniform in the column and the velocity of the droplet is constant during its free movement and it is possible to eliminate the end effects. The mass exchanged during the free movement of the droplet can be calculated. To calculate the mass transfer coefficient, one needs the concentration vs time course, which unfortunately is unknown. Therefore, assumptions have to be made. Because of the complex hydrodynamcis of the droplets, it is difficult to make an accurate estimation.

To avoid these difficulties and to carry out differential measurements for very short observation times (.e.g for measurements during droplet formation), a modified liquid scintillation technique was developed to investigate mass transfer on freely suspended droplets. The development of this technique started in 1964, and since that time it has been improved [12, 74, 49, 88, 89].

The droplet consisting of solvent (e.g. xylene) or extractant + diluent (e.g. amine + xylene) and solute (e.g. acetic acid) as well as the scintillators (PPO + POPO) are irradiated by low energy γ-rays ($E_{max} = 60$ keV) by using a closed Am 241 source with a well-focused ray (Fig. 2.19). As a consequence of the irradiation of the solvent, low energy electrons are formed by the photo- and/or Compton effect with an energy spectrum similar to that of tritium. They excite the solvent molecules, which transfer their energy to the primary and then to the secondary scintillators, which emit this energy as fluorescent light at 300–400 nm (Fig. 2.20). Hence, the external γ-source acts like a solvent label. The advantage of this labelling is that no open radioactive sources (radio-nuclides with low energy β-activity) are needed. This energy transfer mechanism is disturbed in the presence of solutes with polar groups (chemical quench [90]). Therefore, if the solute is extracted from the droplet, the count rate increases and, if it is transferred into the droplet, the count rate diminishes as time goes on.

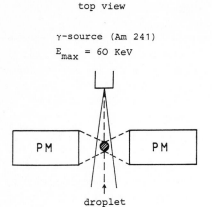

top view

γ-source (Am 241)

E_{max} = 60 KeV

PM   PM

droplet

consisting of solvent (toluene, xylene)

solute (salicylic acid, phenylalanine)

scintillator (PPO + POPOP)

Fig. 2.19. Measuring setup for using the modified liquid scintillation technique for contactless determination of the solute concentration in freely suspended liquid droplets [89]

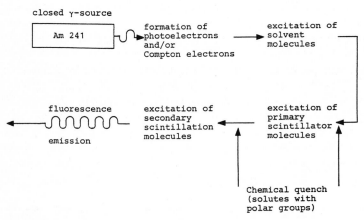

closed γ-source

Am 241 → formation of photoelectrons and/or Compton electrons → excitation of solvent molecules

fluorescence emission ← excitation of secondary scintillation molecules ← excitation of primary scintillator molecules

Chemical quench (solutes with polar groups)

Fig. 2.20. Energy transfer in the droplet [89]

A comparison of pulse height density distributions in absence and in presence of chemical quench indicates that in the latter case the distribution shifts to lower energies (pulse heights) (Fig. 2.21). This change of fluorescence emission is measured by two photomutlipliers (Fig. 2.19). The signals are processed according to Fig. 2.22. The coincidence pulses of the two multipliers are counted in the consecutive channels usually at equidistant time periods of 1 s. However, for measuring the mass transfer during droplet formation equidistant time periods of 10 ms were used [91]. Usually each run consists of 1000 to 2000 measured points. The measured pulse rate vs the time curves are smoothed by a special service routine using.50 to 250 points. The variation of the solute amount in the

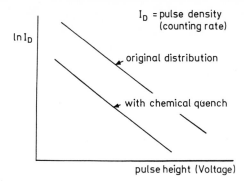

I_D = pulse density (counting rate)

$\ln I_D$

original distribution

with chemical quench

pulse height (Voltage)

Fig. 2.21. Pulse height density distribution [89]

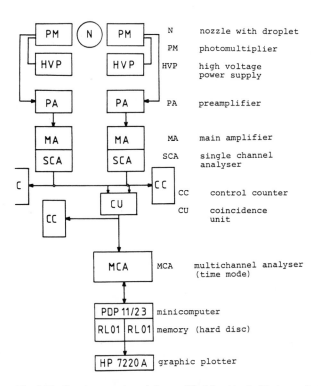

N	nozzle with droplet
PM	photomultiplier
HVP	high voltage power supply
PA	preamplifier
MA	main amplifier
SCA	single channel analyser
CC	control counter
CU	coincidence unit
MCA	multichannel analyser (time mode)
PDP 11/23	minicomputer
RL01 RL01	memory (hard disc)
HP 7220A	graphic plotter

Fig. 2.22. Signal processing of the modified liquid scintillation technique [89]

droplet is calculated by a calibration curve and the solute concentration by the known droplet volume. The measuring section consists of a quartz nozzle with stainless steel grids located upstream from it and stainless steel hypodermic needles for the simultaneous formation of one, two or three droplets in the nozzle. The droplet formation rates and sizes are controlled by automatic burettes. In the plane of the freely suspended droplet a closed Am 241 source is mounted in the nozzle holder (Fig. 2.23) and the two photomultipliers are optically coupled to the nozzle wall (not shown in Fig. 2.23). The droplet is held at rest in the nozzle at a definite height by means of a flat velocity profile of the downstreaming liquid, forming a velocity minimum at the nozzle axis. The drag force is balanced perfectly by the buoyancy force of the solvent droplet. With this setup, the solute concentration in the droplet, consisting of nonquenching solvent can be measured as a function of time for extended periods of time (many hours) with fixed and/or freely suspended single droplets at different Reynolds numbers (in a laminar and transient range as well as in a pulsed flow of the continuous phase).

This technique yields high precision mass transfer data with high time resolution. As it measures the solute amount by an optical method, it does not

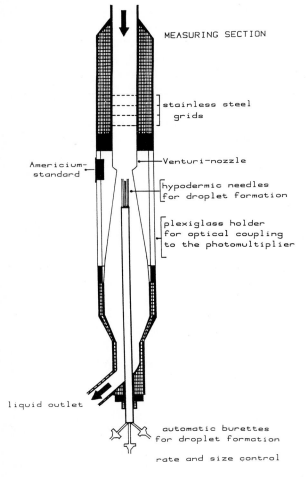

Fig. 2.23. Measuring section of the modified liquid scintillation setup [89]

need the problematic sampling procedure of the other experimental techniques. The disadvantage of the technique – that only nonpolar solvents can be used – should be overcome in the future by increasing the quantum yield by improved equipment construction.

This highly sophisticated technique needs an operator with experience and high precision equipment. This is probably the reason why other research groups have not yet established this technique.

2.4.3 Residence Time Distributions

In Chapter 2.1.5, system models are presented. For the identification of the model parameters, residence time distribution RTD measurements of the continuous phase are usually carried out.

The parameters of the cascade model are the mean residence time τ and the number of ideal stages N in the cascade. The parameters of the dispersion model are the mean residence time τ and the axial dispersion coefficient D_{ax}.

In biotechnology, the product is usually recovered from the aqueous broth, which is the continuous phase. Therefore, the parameters of the cascade model are τ_a and N_a, and those of the dispersion model τ_a and D_{axa}. The determination of the RTD is a well-established stimulus/response technique in chemical reaction engineering [92].

The general relationship between test function x(t) and response function, y(t), is given for a linear lumped parameters system by the convolution integral:

$$y(t) = \int_0^t x(t - t')\,g(t)\,dt',\qquad(2.96)$$

where g(t) is the weighting function of the investigated system.

If the test function is an ideal δ-function: $x(t - t') = \delta(t)$, the response function is identical to the weighting function: $y(t) = g(t)$, and it is called the 'residence time density function'.

If the test function is an ideal step function, the response function is the integral of the weighting function and it is called the 'residence time function' or 'residence time distribution'.

A great many papers have been published in recent years on RTD [93]. Dye and salt solutions are used as tracers in the aqueous phase and dye solutions in the organic phase. Nonideal delta functions (pulses) or nonideal step functions of tracers are frequently used as test functions. Recently, pseudostochastic test signals have become popular.

The response function is usually measured by an electrical conductivity probe (salt tracer) or photometer (dye tracer). Since no ideal delta or step function can be realised, Eq. (2.96) has to be solved for g(t). Since all of the functions in Eq. (2.96) are linear, it is transformed to eliminate the convolution integral:

$$Y(p) = X(p)\,G(p),\qquad(2.97a)$$

where $Y(p)$, $X(p)$ and $G(p)$ are the transform of $y(t)$, $x(t)$ and $g(t)$ and $p = i\omega$ for Fourier transformation and $p = a + i\omega$ for the Laplace transformation. The transformed weighting function (Fourier transform of $g(t)$ is the frequency response function, and the Laplace transform of $g(t)$ is the transfer function) is calculated by

$$G(p) = Y(p)/X(p). \tag{2.97b}$$

The back transformation yields the weighting function $g(t)$. The transformation of the functions can be carried out during the measurement in real-time by fast Fourier transformation (FFT) computer.

The identification of the model parameters are performed by comparing the calculated and measured weighting functions or their transforms.

Another, less exact technique uses the moments of the test and response functions. The relationships between them are:

$$\mu'_{y1} = \mu'_{x1} + \mu'_{g1} \tag{2.98a}$$

for the first moments above the origin and

$$\mu_{y2} = \mu_{x2} + \mu_{g2} \tag{2.98b}$$

for the second moments above the first moment,

where μ'_{y1}, μ'_{x1}, μ'_{g1} are the first moments of $y(t)$, $x(t)$ and $g(t)$ and μ_{y2}, μ_{x2}, μ_{g2} are the second moments of $y(t)$, $x(t)$ and $g(t)$ above the corresponding first moments.

The moments of the weighting functions are given by

$$\mu'_{g1} = \mu'_{y1} - \mu'_{x1} \tag{2.99a}$$

$$\mu'_{g2} = \mu'_{y2} - \mu'_{x2}. \tag{2.99b}$$

For the parameters of the cascade model (Chapter 2.1.5)

$$\mu'_{g1} = \tau \tag{2.100a}$$

and

$$\frac{\mu_{g2}}{\mu'^2_{g1}} = \frac{1}{N} \tag{2.100b}$$

hold true.

For the parameters of the dispersion model with "open" system boundaries [94, 95]

$$\mu'_{g1} = \tau \tag{2.101a}$$

and

$$\frac{\mu_{g2}}{\mu'^2_{g1}} = \frac{2}{Bo} \tag{2.101b}$$

and with a "closed" system boundary at the entrance and "open" boundary at the exit [94, 95]

$$\mu'^2_{g1} = \tau(1 + \text{Bo}^{-1}) \tag{2.102a}$$

and

$$\frac{\mu_{g2}}{\mu'^2_{g1}} = \frac{2}{\text{Bo}} + \frac{3}{\text{Bo}^2} \tag{2.102b}$$

are valid.

Here $\text{Bo} = uH/D_{ax}$ Bodenstein-number, H is the length of the active extraction section, D_{ax} is the axial dispersion coefficient, $u = H/\tau$ is the actual liquid velocity.

2.4.4 Symbols for Sect. 2.4

Bo	$= uL/D_{ax}$ Bodenstein-number	1
D_{ax}	axial dispersion coefficient	$L^2 T^{-1}$
G(p)	transform of g(t)	
g(t)	weighting function	
H	length of the active section of the extractor	L
N	number of stages in the cascade	1
u	actual liquid velocity	$L T^{-1}$
X(p)	transform of x(t)	
X(t)	test function	
Y(p)	transform of y(t)	
y(t)	response function	
μ'_1	first moment above the origin	T
μ_2	second moment above the first moment	T^2
τ	mean residence time	T

3 Extraction Equipment

In the chemical industry, three main types of extraction equipment are in use:

- mixer-settler extractors
- extraction columns and
- centrifugal extraction equipment.

In biotechnology practice, however, few of them are used. According to the literature [102–106], mixer-settlers are applied in a broad field of the chemical industry, but not in biotechnology, and for this reason mixer-settlers are not considered here.

The centrifugal extraction equipment is most popular in antibiotics production, where large broth volumes have to be handled, which sometimes form stable emulsions with the solvent. The use of extraction columns in biotechnology is limited to a few extractor types [107–117]. Their selection and scale-up in biotechnology has not yet been dealt with [118–119].

3.1 Centrifugal Extraction Equipment

Centrifugal extraction equipment was developed for biotechnological practice, and that still is its main application. Excellent reviews are available on this equipment [96–98], some of the articles treat their specific use in biotechnology [99–101]. This equipment is usually operated in a coutercurrent mode [97–101, 129–132, 136]. It is especially suitable for handling solvent and raffinate phases with small density differences.

Two main types of extraction equipment can be distinguished:

- centrifugal extractors and
- centrifugal separators.

Centrifugal separators are analogous to mixer-settlers, as in each of their stages a complete phase separation occurs. They are mixer-settlers, where in the settler the phase separation is accelerated with centrifugal force. Some of this equipment consists of separate mixer and settler units. In others these units are combined.

In centrifugal extractors (differential contact centrifugal extractors), no discrete stages can be recognized. They consist e.g., of several concentric perforated cylinders and work as "rotating perforated plate columns". This concept of using centrifugal force to achieve countercurrent mixing and separation was developed in the early 1930s. The principle of this design is to introduce the heavy phase close to the axis of a rotating container, while the light phase is introduced close to the periphery. The applied centrifugal force causes the two phases to move radially in a countercurrent mode in such a way that the heavy phase is moved to the periphery and the light phase to the rotation axis [97]. The first construction of this type was developed by Placek [133] and Podbielniak [134] with a horizontal axis and later on with provision for a phase introduction and removal through an incorporation of radial tubes [135]. By 1944 this extractor was successfully tested for penicillin extraction [99].

Figure 3.1 shows the most popular equipment for penicillin recovery, the Podbielniak centrifugal extractor of Baker Perkins Chemical Machinery Inc. [136]. The introduction and removal of the phases is provided through the axis. To do this, 4 to 7 bar prepressure is necessary. At throughputs of 130 m^3/h, three to six and one half theoretical stages can be attained. The density difference of the phases must be larger than 50 kg/m^3. Hydrodynamically balanced mechanical seals ensure hermetic operation. The position of the interface of the phases is controlled by their back pressure. For each special extraction system the built-ins (number and distance of the central cylinders and the hole geometry) have to be optimized. This is usually carried out in small laboratory equipment (A1) with a volume of 0.6 l [98].

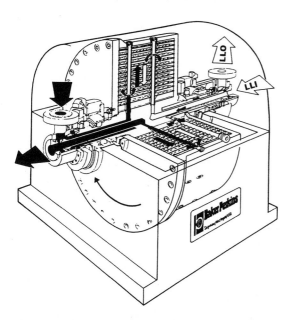

Fig. 3.1. Podbielnak centrifugal extractor (Baker Perkins, Michigan)

According to Anderson and Lau [101] the Podbielniak Extractor Type 9500 is operated at 2700 rpm, $R = Q_h/Q_l = 3.4-4.5$, $Q_t = 7.5 \text{ m}^3/\text{h}$ with 0.42–0.57 theoretical stages and 60–70% extraction efficiency, where Q_h means heavy phase, Q_l light phase and Q_t throughput. Since these data are fairly old, one can assume that the performance of this extractor has been improved considerably by now.

The Quadronic Extractor is also a horizontal countercurrent extractor (Fig. 3.2). The working principle and the introduction and removal of the phases are similar to that of the Podbielniak Extractor. The quadronic Extractor is easier to dismantle than the Podbielniak equipment, when the introduction and removal of the phases at optimal radial positions is to be facilitated. The controlled mixing is achieved by using different orifice sizes at different radial positions. Three different internal designs are used:

– the orifice disc column
– the perforated strips and
– sectional concentric orificed bands [132].

The mixing intensity in a particular radial position depends on the construction of the built-ins. As a special variation of the mixing intensity considerably influences the performance of the extractor, it is recommended that the built-ins best suited to the system to be separated are used [119]. Four to six theoretical stages can be attained at the throughput of 150 m³/h [132].

In contrast to the Podbielniak and Quadronic Extractors the Alfa-Laval differential contact centrifugal extractor ABE 216 (Fig. 3.3) has a vertically rotating bowl. Both phases enter under pressure at the bottom of the extractor. The light phase enters near the periphery and the heavy phase near the center. The two phases move radially and countercurrently and mix as they enter or leave each channel through the joining orificed openings. The heavy phase leaves at the top by way of the outermost channel. The interface is established in every channel, and the two phases flow in countercurrent layers except at the

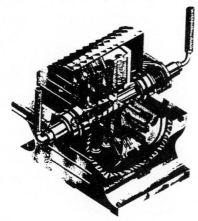

Fig. 3.2. Quadronic centrifugal extractor liquid dynamics, Illinois bulletin No. 15, 1968

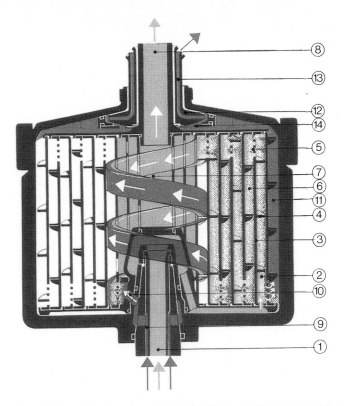

Fig. 3.3. Centrifugal extractor ABE 216 (Alfa Laval PP60926 T). (*1*) feed, (*2*) extraction channel, (*3*) insert, (*4*) helical compartment, (*5*) openings, (*6*) outlet of the light phase, (*8*) extractor drum (*9*) inlet of the heavy phase, (*10*) passage of the heavy phase, (*11*) clarification chamber, (*12*) barking disc, (*13*) outlet of the heavy phase, (*14*) passage of the heavy phase to the outlet; light phase (*light field*), heavy phase (*dark field*)

mixing ports [97]. According to the Alfa-Laval prospect [130], the following operation conditions are typical for a penicillin extraction in ABE 216: the throughput and extraction efficiency vary from plant to plant depending on nutrients used, conditions of the fermentation, etc., but normally the throughput is in the range from 5 to 9 m³/h of the filtered broth. The ratio of broth/solvent is normally in the range from 3:1 to 5:1, but also ratios of up to 8.1 with a penicillin recovery of 96–98% have been reported.

The traditional way is to perform the extraction in three steps, where the penicillin – in the enriched solvent after the first extraction step – is extracted into a buffer phase. The enriched penicillin from the second step is then extracted into a solvent phase in a third step.

The centrifugal separators: Luwesta Extractor [129] and Westfalia TA Extractors [129] work according to the mixer-settler principle, where the settling is accelerated by centrifugal force. They are suitable for the separation of solvent and raffinate phases with a low density difference ($\Delta\rho > 20$ kg/m³).

The single stage Westfalia TA is a disc-type centrifuge with a stationary bowl and rotating shaft directly mounted on the driving engine (Fig. 3.4). Both phases enter and leave the extractor at the top [129] Alfa-Laval also offers disc separators.

These extractors consist of a mixing chamber and a separator in which a large number of conical discs are positioned. Colinear holes on the discs in a given radial distance from the center form vertical channels, through which the liquid mixture flows upward. During its passage through the channel, the mixture is separated: the heavy phase moves outward and the light phase inward. The interface of the phases is at the radius, where the center of the channels are positioned. The optimal radial position of the interface and the channel center depends on the system to be separated. The position of the interface is controlled by the position of the heavy phase outlet by means of a suitable diameter of the heavy phase collector ring. The exact positioning is obtained by back pressure control.

The Luvesta Extractor consists of a cascade of mixers and settlers with countercurrent flow of the two phases. The throughput of the single stage extractors vary from $1.25 \, m^3/h$ (TA 7-04) to $120 \, m^3/h$ (TA 275-04).

The advantage of the centrifugal extractors and separators are

– their extremly high throughput
– very short contact time (few seconds), which is important for the extraction of unstable solutes (e.g., Penicillin G at PH 2)
– low solvent volume in the extractor.

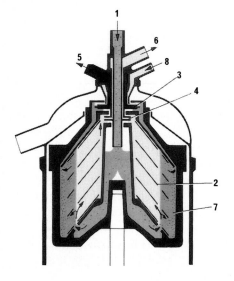

1 Feed
2 Discs
3 Centripetal pump for heavy phase liquid and admixed light liquid
4 Light phase centripetal pump
5 Heavy phase and admixed light phase liquid
6 Light phase
7 Sediment holding space
8 to mixing chamber

Fig. 3.4. Single-stage extractor type TA 40-40 (Westfalia, Oelde)

Their disadvantages are

- high investment costs
- high variable costs
- high pressures (up to 100 bar) can occur locally at the periphery of the rotating cylinder. (The phase diagram and the equilibrium curve are influenced by the pressure variation which has to be taken into account for the calculation of the necessary separation stages).

The other differential contact centrifugal extractors (unpressurized vertical UPV extractor) and centrifugal separators (Robatel SGN and BXP Extractors, SRL and ANL as well as MEAB SMCS 10 extractors) have not yet been used in biotechnological production processes [97].

Recently Westfalia has developed an Extractor-Decanter which is used for the direct extraction of antibiotics from fermentation broths and mycels without their being separated (Fig. 3.5) [129]. The two phases are fed into the rotating bowl through separate inlets at different parts of the bowl. Phase contact is effected in a countercurrent flow in the feed area and in the spirals of the scroll. The separation zone is adjusted for variable liquid densities by means of the ring dam at the cylindrical bowl end. After mass transfer into the contact zone, the enriched solvent (light phase) is discharged at the cylindrical bowl end under pressure by a centripetal pump.

Depending on the design of the scroll, it is possible to use solvents whose specific gravity is greater or less than that of the suspensions which are to be extracted. The raffinate phase (suspension) discharges at the conical end of bowl. Solid particles which are sedimented in the bowl are conveyed by the scroll to the conical end of the bowl and discharged together with the suspension [129].

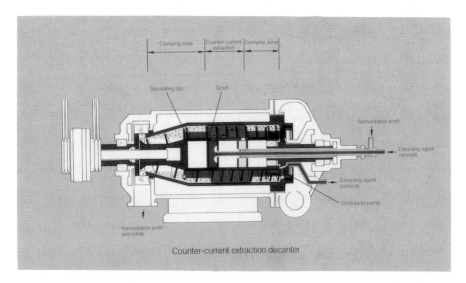

Fig. 3.5. Counter-current extraction decanter (Westfalia, Oelde)

The drum of the extraction decanter CA-226-120 (with 220 mm in diameter and 750 mm length) is operated at 5000 rpm and with a difference of 32 rpm between rotational speeds of the drum and the scroll. An overall throughput of 1.3 to 2 m³/h was obtained with an antibiotics-yield of 97–98%.

3.2 Extraction Columns

Nonmechanically agitated contactors, such as spray columns, baffle contactors, packed columns and perforated plate columns are usually not used in biotechnology [107]. Of the mechanically agitated extraction columns, only bench- and pilot plant-scale

– Kühni columns [115]
– Karr columns [110] and
– pulsated perforated columns [109]

have been used in biotechnology. Other mechanically agitated columns (pulsed packed columns [108], rotating disc contactors [111], asymmetric rotation disc extractors [112], Oldshue-Rushton columns [114], Scheibel columns [113] and others [117] are used in the chemical and related industries only.

The RTL (Graesser raining-bucket) contactor [116] was developed for low density differences in solvent and raffinate phases and for low interfacial tension systems. Therefore they are suitable to handle aqueous two phase systems which are used in the recovery of proteins (e.g. enzymes) from disrupted cells. However, no industrial application of this extractor is known.

The Kühni column (Fig. 3.6) consists of an assembly of turbine mixers mounted on a central shaft. The turbine mixer is a double entry radial flow

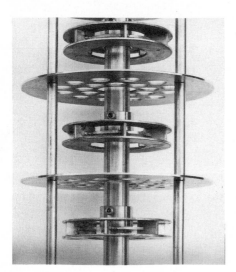

Fig. 3.6. Kühni contactor Impeller and stator (Kühni, Switzerland)

impeller which generates a characteristic flow pattern of the Kühni-contactor compartments (Fig. 3.7). The column also consists of stator plates which separate the compartments. They are provided with circular holes to allow the axial flow of the two phases through the column [115]. By variation of the geometry of the compartments, the size of the mixing turbine and the available free area through the stator plates, the performance of the column can be changed in a broad range. This flexibility allows one to adapt the column properties (axial mixing, holdup of the dispersed phase, droplet-size distribution) to the particular system to be separated. Columns from 6 cm to 2.5 m in diameter are used in various (petrochemical, heavy chemical, hydrometallurgical and pharmaceutical) industries as well as for waste water treatment.

Only a few measurements were carried out to determine column throughput, flooding, drop size and holdup, axial mixing as well as mass transfer rate as a function of the column construction and operation mode [115].

The Karr columns, belonging to the reciprocating-plate extraction columns, attained the widest use in the industry of western countries. This column has plates with large holes and a large free area ($\approx 58\%$). Another type has plates with small holes and a small free area with or without downcomers. The latter type is manufactured commercially under the trademark 'vibrating plate extractor (VPE)' and is in use in eastern Europe and the former Soviet Union. There are two modifications of VPE: columns with parallel motion of the plates and those with counter motion of the plates [138, 139]. Both Karr and VPE extractors are used in biotechnology [110].

The column developed by Karr and Lo consists of a stack of perforated plates and baffle plates that have a free area of about 58%. The central shaft that supports the plates is reciprocated by means of a reciprocating drive mechanism located at the top of the column (Fig. 3.8). The amplitude is adjustable generally in the range of 3 to 50 mm, and the reciprocating speed is variable up to 1000 strokes/min [110]. Also baffle plates are used periodically in the stack to minimize the axial mixing.

Baird et al. [140–144] investigated the hydrodynamics (flooding, drop size, holdup, power consumption and axial mixing) of the Karr columns. Recommendations for the design as well as far the calculation of the performance and scale-up of Karr columns are given in [145–147]. The advantages of the Karr columns are [110].:

– high throughput and high mass transfer as well as high volumetric efficiency,
– high degree of versatility and flexibility,
– easy handling of emulsifiable material and liquids with suspended solids.

The vibrating-plate extractors (VPE) are built with a parallel plate motion and with a counter motion of the plates. In the former, a stack of perforated plates are supported by a central shaft which moves back and forth by means of a reciprocating drive mechanism, similar to that of the Karr column. The plates have small circular holes for the dispersed phase and one or more large openings for the continuous phase. The separation of the passages for continuous and

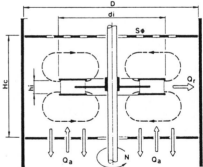

Fig. 3.7. Kühni contactor *Top* column; *Bottom* characteristic flow pattern and geometry of the Kühni contactor

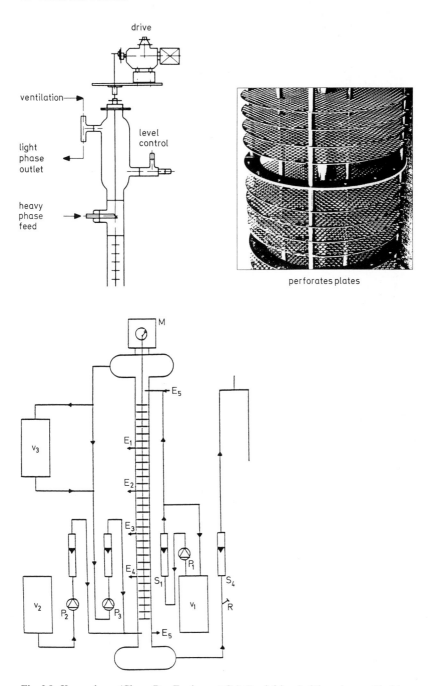

Fig. 3.8. Karr column (Chem. Pro-Equipment Co). *Top left* head of the column with drive; *top right* perforated plate and baffle assambly *bottom* entire equipment

dispersed phases ensures high throughputs of the dispersed phase even in case of high flow rates of the continuous phase. It also permits the maintenance of a stable mixer-settler regime over a wide range of plate velocity (amplitude × frequency). In the small perforations, drops are formed and reshaped by a mechanism similar to that of a periodical outflow from the nozzle. The dispersion of the drops does not require high plate velocity, thus VPE operates at relatively low amplitudes (0.3 cm) and frequencies (120 to 300 per min). Consequently, mechanical stress and energy consumption are low [110].

With a counter motion of the plates, the geometry of the column is quite similar to that with a parallel plate motion. However, the set of plates is divided into interlaced stacks, each of which performs its own harmonic motion. These motions are of the same amplitude and frequency, but are shifted in the phase by 180°. The group of plates (1) is fastened to the shaft (2) actuated by the excentric disc (3). The other group of plates (4) is fastened to the shaft (5) and actuated by the excentric disc (6) (Fig. 3.9). The countermotion of the plates ensures high mechanical stability of the apparatus and high extraction efficiency [110].

Pulsed perforated plate columns (with fixed plates and pulsed liquids) became the classical extractor for nuclear reprocessing, but they are also used in biotechnology practice [110].

A series of horizontal plates are fixed in the column, and the liquid is pulsed by a reciprocating motion of bellows or a piston in a cylinder. The two phases flow countercurrently. By the pulsing action applied at the base of the column the dispersed light phase is forced through the perforations of the plates on the upstroke, and the continuous heavy phase flows downward on the downstroke. If the heavy phase is to be dispersed, it is forced through the holes on the downstroke, and the continuous light phase flows upward on the upstroke [109]

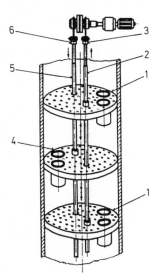

Fig. 3.9. VPE column with counter motion of the plates [110]

(Fig. 3.10). Typical frequencies are in the range of 60 to 180 per min and amplitudes of up to one plate spacing are used. The different regions of operation are shown in Fig. 3.11. The sum of flows of both phases is plotted against the plate velocity (frequency × amplitude). At low pulsation and flow rates the dispersed phase coalesces under (or above) the plate (mixer-settler region). Flooding occurs if the phases are fed into the column at a greater rate

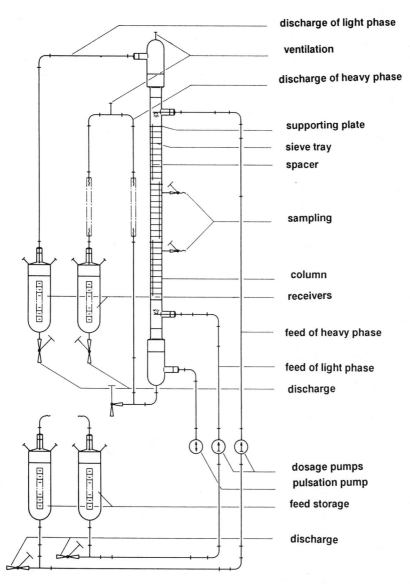

discharge of light phase

ventilation

discharge of heavy phase

supporting plate

sieve tray

spacer

sampling

column

receivers

feed of heavy phase

feed of light phase

discharge

dosage pumps
pulsation pump
feed storage

discharge

Fig. 3.10. Pulsated perforated plate column

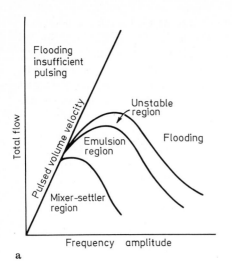

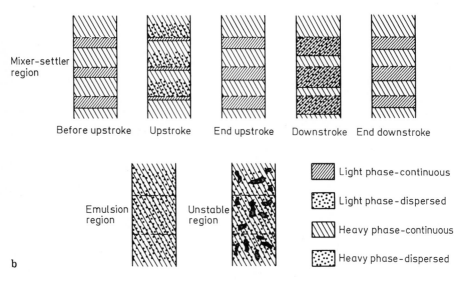

Fig. 3.11.a, b Pulsated plate column operating regions: **a** total flow versus frequency × amplitude; **b** phase dispersion [109]

than they can usually pass through the plates just by the pumping action of the pulser. In this case, each of the phases is discharged at the end of the column at which it enters. This leads to poor mass transfer rates and low column performance.

As the pulsing and flow rates increase, the inertial and shear forces increase. These forces hinder the coalescence of the drops, and dispersion or emulsion is formed. The greater the pulsing (specific power input) the smaller are the drops. At constant flow rates with increasing plate velocity, flooding can also occur. In this region, the highest mass transfer rates are obtained. Between emulsion and flooding regions, the emulsion becomes locally unstable and coalescence and phase inversion occurs [109]. "Standard geometry of plates" are recommended for the nuclear industry: plate thickness up to 2 mm, holes 3 mm in diameter on a triangular array, spaced 50 mm apart (23% free area) [148]. However, it is not known if this geometry also applies to fermentation media.

In handbooks and publications on extraction quantitative relationships are frequently recommended for design and operation of the equipment. However, all relationships were evaluated by model systems which are dissimilar to fermentation media.

The influence of their protein content in particular is unknown. Therefore, the recommended relationships cannot be used for biotechnological purposes. The extraction equipment dealt with here has to be optimized empirically with the particular fermentation broths.

4 Extraction of Metabolites

One usually distinguishes between primary and secondary metabolites. Primary metabolites are simple compounds which have direct connections to the metabolic pathway of the microorganisms. Therefore, close relationships exist between the growth rate and their production rate.

Secondary metabolites are usually complex compounds. Their biosynthesis has no direct connection to the cell metabolism. Consequently, there is usually no close relationship between cell growth and secondary metabolites production.

Typical primary metabolites are ethanol, acetic acid, citric acid and amino acids.

Typical secondary metabolites are antibiotics and vitamins. Both types of metabolites are excreted from the cells into the fermentation broth.

Downstream processing of primary metabolites usually consists of only a few steps: recovery from the broth combined with the enrichment and purification of the raw product.

The concentration of secondary metabolites in the fermentation broth is usually much lower than those of the primary metabolites. Their recovery and enrichment may consist of several steps. Since most of them are used as therapeutics, the quality requirements of the products are high. Therefore, for their purification numerous steps are necessary.

The extraction can be one of these recovery and/or purification processes.

Several products are manufactured by microbial biotransformation. They can also be recovered and/or purified by extraction. Many of them are hydrophobic (e.g., steroids), which are extracted by hydrophobic solvents (methylene chloride, ethylene chloride, chloroform). They are not considered here.

4.1 Recovery of Alcohols from Fermentation Broths

Research of fermentative production of alcohols from renewable sources (plant biomass) and from agricultural wastes has been intensified during the last decade for their use as possible energy sources, such as their addition to gasoline or diesel oil.

Since the recovery of alcohols from fermentation broths is an expensive process, several research groups have been trying to find an inexpensive way.

One of the possibilities for recovery seemed to be liquid extraction. The necessary data have been worked out all over the world.

Recent results on the extraction chemistry of low molecular weight aliphatic alcohols was reviewed by Kertes and King in their comprehensive report ([5] in Chapter 2). This report discusses physical and thermodynamic properties, solubilities and partition of alcohols in binary and ternary systems, furthermore, their extraction by nonpolar and polar solvents, the coextraction of water, the effect of salting out agents and temperature on their partition coefficients.

Of these alcohols, the primary metabolites ethanol and butanol are of main practical interest. In the chemical industry, these alcohols are petrochemical products. They are formed in high concentrations and purified by means of fractional distillation [149, 150].

Fermentative production of ethanol is of importance for edible purposes [151]. Because of the product inhibition of yeast or bacterial cell growth and product formation the ethanol concentration in fermentation broths are always low, usually from about 5 to 10%. In industrial practice, this ethanol is enriched by fractional distillation. For the production of absolute ethanol, azeotropic distillation is used. Because of the high specific energy consumption of the distillation, several research groups have been on the lookout for alternative techniques; however, up to now, no better technique has been found [152].

The loading capacity of adsorbents is too low (e.g., [153]), pervaporation does not work at low ethanol concentration [144] and the solvent extraction is not yet economical [155]: Essien and Pyle [184] investigated different solvents for ethanol recovery with regard to their solvent inventory, solvent make up and energy costs. Heptanal was found the most economic solvent. A comparison of the economy of solvent extraction and distillation was carried out based on the separation of $120 \, m^3 day^{-1}$ of 92.4% ethanol from aqueous feeds of 5, 9.6 and 14 wt% ethanol with 98.5% ethanol recovery $24 \, h \, day^{-1}$ and 330 days year^{-1} operation.

Solvent loss contributes more than 50% for feeds below and around 12 wt% and increases rapidly with decreasing concentration. Capital related costs range from around 16% of the total to 20% as a feed concentration increases from 5 to 14 wt%. Capital-related costs for solvent extraction are about 60% higher than for distillation. This more than offsets any advantage in energy and utility [184].

Since the investigations indicate that extraction cannot compete with fractional distillation, the research is concentrating on the extractive fermentation, i.e., recovery of alcohol by solvent extraction from fermentation broth during alcohol formation. In this field, e.g. the extraction of butanol competes with the perevaporation technique [159]. The advantage of the removal of alcohol from the fermentation broth during its formation is the increase of the productivity due to decrease of product inhibition.

The main problem is to find the suitable solvent with a high distribution coefficient, especially for ethanol, which does not form a stable emulsion with the fermentation broth, is easy to recover from the extraction phase and does not have a toxic effect on cells or does not cause inhibition, i.e, it has good biocompatibilty [155, 156, 160–162].

The following solvents have been investigated and recommended: dodecanol [156, 163–165], higher iso-alcohols [156, 166], higher n-alcohols [155, 156, 168], tributylphosphate [165, 166, 172], dibutylphtalate [157], dodecane [169], fluorocarbons [170], higher iso-acids [169], and aqueous two-phase systems [171]. Dodecanol is, e.g., fairly biocompatible, however, it tends to form a stable emulsion with fermentation broths, it has a relatively high solubility in water, its melting point is 26 °C, which is too high for some fermentation processes, and the ethanol distribution coefficient in a dodecanol/water system is only about 0.35 on the mass basis.

One can use pure solvents or one can dilute the solvents by, e.g., iso-paraffins. Such a blend is the tri-n-butylphosphate-ISOPAR M-systems [172, 173]. The fermentation broth was extracted directly with this blend in a Karr column at 50 °C, and the raffinate was accumulated and recycled into the fermenter after cooling it back down to the fermentation temperature. At an initial concentration of ethanol of 10 wt%, the extract composition ranged from 93 to 97 wt% while the product (ethanol) was recovered up to 70 to 75 wt%, on the solvent free basis. This reduction of ethanol content may be due to vaporisation losses of ethanol during extraction. Because none of these investigated solvents gave results which could yield an economical process, a large number of solvents were examined by a computer to find the most promising ones.

Kollerup and Daugulis [160–162] investigated 1500 solvents by a computer program regarding their suitability as extractants and, based on this screening, 62 solvents have been tested with yeast cultures to determine their biocompatibility. Fifteen have proven to be completely biocompatible. The main problem of the extractive fermentation is that solvents with high distribution coefficients for ethanol are toxic for the cells and solvents which are biocompatible have low distribution coefficients. Matsumura and Märkl [174] recommended an interesting solution to this problem by using a barrier for the solvent molecules beneath the surface of gel beads, immobilizing the cells as a protection against solvent toxicity. Poropack Q was found to be an effective barrier, and the ethanol production rate of immobilized cells protected with Poropack Q did not change even after the production of eight batches in fermentation broth, saturated with sec-octanol, which is a very toxic solvent for the used yeast and bacteria [174]. Different vegetable oils also have protecting effects on cells immobilized in Ca-alginate [185].

Supercritical fluid extraction was also considered for the recovery of ethanol from fermentation broths. Generally, supercritical fluids are suitable for the extraction of lypophilic compounds of low polarity, like esters, ethers, lactones below 70 to 100 bars.

Introduction of strong polar groups (e.g., –OH, –COOH) makes extraction difficult. Benzene derivatives up to three –OH, or two –OH and one –COOH groups can still be extracted, but above 100 bars. Benzene derivatives with three –OH groups and one –COOH group, as well as sugars and amino acids cannot be extracted below 500 bars. In the industry, supercritical extraction is used in the food as well as in the mineral oil industry.

For the extraction of ethanol from aqueous solutions, supercritical carbon dioxide [176–178], ethane [178] and propene [175] were used.

Figure 4.1 shows the process scheme of supercritical gas extraction. The feed contains components A (ethanol) and B (water). In the separator stage, the solvent (supercritical gas) is added to the feed. At a high selectivity of the solvent, for A, component A is enriched in the solvent (P1) and component B in the residue (P2). By reducing the pressure and/or increasing the temperature in the solvent regenerator unit, the solute solubility is reduced and component A is recovered.

In the triangular diagram, the binodal curve separates the two phase regime from the single-phase one, as in the case of the classical extraction. Point C represents the composition of the feed + solvent, which decomposes into the equilibrium mixtures D (solvent dissolved in feed) and E (components A and B dissolved in the solvent). With the separation of these equilibrium phases, products P_2 and P_1 are obtained. The solubility of the solute in supercritical gases depends on their density, which can be varied by means of operational pressure and temperature in broad ranges. Close to the critical temperature, small pressure and temperature variations cause large density variations. This allows complete separation of the solute from the solvent in product P1. Close

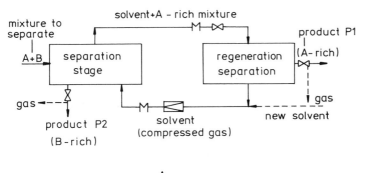

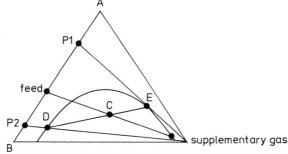

Fig. 4.1. Process scheme of supercritical gas extraction [178]

to the critical point, the vaporization enthalpy of solvent is also low. This allows a simple separation between solute and solvent, which only has a low energy requirement. Also, the transfer rates are high due to low viscosity, high diffusivity and heat conductivity.

The solvents (CO_2, ethane, ethane etc.) are completely miscible with ethanol, and their water solubility is low. The separation factors of ethanol in supercritical CO_2 ($T_{crit} = 30\,°C$, $P_{crit} = 74$ bars) are given in Fig. 4.2.

The separation factors for the CO_2-solvent are considerable: two to seven times higher at low ethanol concentration than those of the binary system, but with an increasing ethanol concentration this difference reduces (Fig. 4.3).

The solubility of ethanol in supercritical CO_2 increases with an increasing ethanol concentration in the aqueous phase. At 10% ethanol concentration in the aqueous phase, the solubility amounts to 1.5 wt%. Therefore, a large amount of solvent is needed for the ethanol recovery. The solubility of ethanol in supercritical ethane is about the same as that in CO_2. However, its selectivity is by a factor of two larger than for CO_2. Again, with increasing ethanol concentration in the aqueous phase, the separation factor diminishes. For the recovery of ethanol from aqueous solutions with supercritical CO_2 extraction, a 3.4 m high bubble column, 17 mm in diameter, was used [178]. The aqueous ethanol solution was filled into the column and supercritical CO_2 was fed to the bottom of the column, and at its head the gas pressure was reduced and the gas composition was determined. The ethanol concentration in the aqueous phase was measured as a function of time. After 60 min, and at 45 °C, the gas phase

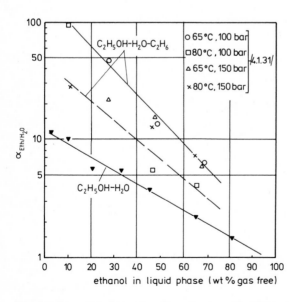

Fig. 4.2. Separation factor for ethanol-water with CO_2 solvent as a function of the ethanol concentration in the water [178]

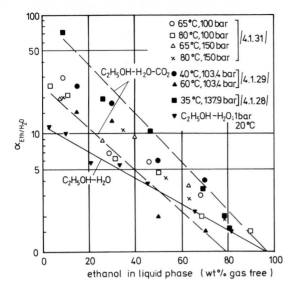

Fig. 4.3. Separation factor for ethanol-water with ethane solvent as a function of the ethanol concentration in water [178]

concentration of 0.8 wt ethanol corresponded to a liquid concentration of 2.3 wt% [178, 179].

In a continuous countercurrent flow of the aqueous and CO_2 phases in this column with 13 wt% ethanol feed the head product contained 51 wt% ethanol (in the CO_2-phase), and the bottom product 0.1 wt% ethanol (in the aqueous phase) at a superficial mean gas velocity of 7.2 mm s^{-1} [178]. The rough estimation of the separation costs indicate that in a 50 000 t/a 90 wt% ethanol plant, about 0.1 US $/kg ethanol could be achieved, which is in the same order of magnitude as the separation costs in a 30 000 t/a ethanol plant [178, 179].

It is expected that by combining different solvent components, the separation factor and solubility of the solute in the solvent can be increased. By addition of solvents to the supercritical gases, the separation factor can be varied within wide ranges [180].

The fermentative production of butanol is considerably hampered by the strong product inhibition [181]. Therefore, the extractive fermentation would considerably improve the economy of the butanol production by *Clostridium acetobutylicum*. Several solvents have been investigated with regard to their use as extractants for acetone and butanol [183]. In Table 4.1 distribution coefficients of acetone and butanol for various solvents are shown. One can observe that distribution coefficients for acetone are always much lower than those for butanol. Alcohols with 6 to 9 carbon atoms are efficient solvents. The increase of temperature and the choice of branched alcohol improves the extraction effici-

Table 4.1. Experimental values of K_D and separation factors α of acetone and butanol for various solvents [183]

Solvents	T(°C)	Acetone K_D	α	Butanol K_D	α
Ethyl ether	20	0.76	51	5.14	412
n-Hexanol	20	0.67	9	9.47	129
	40	1.19	16	14	191
iso-Nonanol	20	0.63	23	7.73	286
	40	0.87	28	8.69	277
iso-Decanol	20	0.50	19	6.88	259
	40	0.66	26	7.87	307
n-Decanol	20	0.41	11	5.86	153
Tetradecanol	40	0.61	19	5.22	160
Diisobutyl-ketone	20	0.72	120	2.55	427
Anisole	20	0.85	643	0.99	750
	40	1.04	334	1.77	570
Phenetole	20	0.68	393	0.88	506
Xylene	20	0.49	806	0.53	878
Mesitylene	20	0.38	1900	0.47	2300
Methylbenzoate	20	0.79	96	1.84	222
Dibutylphthalate	20	0.52	88	1.21	207

ency. Since acetone does not inhibit the growth and metabolite production, its high concentration does not reduce the productivity. When using mixed solvents, acetone can also be removed [220].

Decanol has been proved as a useful solvent [182]. It can easily be separated from the solute butanol due to its high boiling point. The selectivity and partition coefficient of butanol in a decanol/water system is satisfactory. It is toxic for the cells, but it is practically insoluble in the aqueous phase, thus the extraction of the products can be carried out from the cell-free broth. Acetone/butanol was formed in a continuous culture of *Clostridium acetobutylicum*; the productivity was increased by cell retention by means a cross-flow microfiltration membrane module; the butanol was removed by selective extraction from the cell-free broth by decanol, and the butanol-free broth was recycled into the bioreactor. By removing the butanol from the broth, the productivity was improved by a factor of four [182].

4.2 Extraction of Aliphatic Carboxylic Acids

Several aliphatic carboxylic acids are formed during aerobic fermentation as intermediates of the metabolic pathways of the microorganisms. The bioconversion of glucose is performed by the glycolytic Embden-Meyerhof pathway in which pyruvic acid, the key metabolic intermediate, is formed [186]. Pyruvic acid is oxidized in a cyclic manner, known as the tricarboxylic acid pathway, to yield a number of di- and tricarboxylic aliphatic acids of four to six carbon atoms (citric acid, *cis*-aconitic acid, isocitric acid, ketoglutaric acid, succinic acid, fumaric acid, malic acid and oxalacetic acid).

Special modifications of the cycle, such as the glyoxalate bypath, produce some of these acids in relatively high yields. Other modifications involve the reduction of pyruvic acid to lactic acid, or cleavage to yield formic and acetic acids, or propanoic acid as the major endproducts [8]. The three most important fermentation products are acetic, citric and lactic acids. The classical product recovery of acetic acid is a combination of extraction and azeotropic distillation.

Citric and lactic acids are precipitated as Ca-salts from the broth and recovered from their Ca-salts by sulfuric acid. Alternative techniques are based on the extraction. In spite of a number of patents which have been issued, only one process is being used in the industry [201–203].

By specific use of organophosphorus compounds and tertiary aliphatic amines as extractants, the performance of the liquid extraction techniques have been considerably improved. This is the reason why the extraction of organic acids has recently been reinvestigated.

The prerequisite of an economic recovery by extraction is a high partition coefficient and distribution ratio. Several investigations were carried out on the extraction chemistry of carboxylic acids (Tables 4.2). All of these acids are weak as shown by their dissociation constants pK$_a$ (Table 4.3).

The general problems with the extraction chemistry of carboxylic acids were discussed in the excellent review of Kertes and King [9]. They structure their review according to three extraction categories:

– acid extraction by solvation with carbon-bonded oxygen-bearing extractants,
– acid extraction by solvation with phosophorous-bonded oxygen-bearing extractants,
– acid extraction by proton transfer or by ion-pair formation, the extractant being high molecular weight aliphatic amines.

The first two categories involve the solvation of the acid by donor bonds. The distinction between the first two categories is based on the strength of the solvation bonds and the specifity of solvation. Solvation with alcohols, ethers and ketones which are typical carbon-bonded oxygen donor extractants is not regarded as being specific. [9]. The significantly more basic donor properties of the phosphorus-bonded oxygen causes a specific solvation process and the number of solvating molecules per extracted acid is accessible experimentally. In the third category ionic interactions, i.e., reactions are involved.

Under carbon-bonded, oxygen-donor extractants, hydrocarbon and substituted hydrocarbon solvents are taken into account. The Mass Action Law Equilibrium has already been considered for this category under Chapter 2.1.1 (Eqs. (2.1)–(2.33) in Chapter 2). Equation (2.33) in Chapter 2 can also be used for di- and tricarboxylic acids by neglecting the effect of second and third dissociation steps, which are usually smaller than the first by two or three orders of magnitude (Table 4.3). The partition coefficients P$_c$ and dimerisation constants

Table 4.2. Investigation of extraction of aliphatic carboxylic acids with organophosphorus P and amine A extractants

Solute	Extractants	Authors
formic acid	A	Siebenhofer, Marr [193]
acetic acid	P, A	Wardell, King in Chapter 2 [6]
	P, A	Ricker, Michaels, King in Chapter 2
	P, A	Ricker, Pittman, King in Chapter 2
	P, A	King [192]
	A	Siebenhofer, Marr [193]
	A	Wojtech, Mayer [195]
	P	Grinstead [200]
	A	Inoue, Nakashio [197]
propionic acid	P	Pagel et al. [223, 224]
	A	Kawano et al. [216]
butyric acid	A	Siebenhofer, Marr [193]
lactic acid	P	Pagel et al. [223, 224]
	A	Ratchford et al. [210]
	A	Kawano et al. [216, 217]
	A	Pyatnitskii et al. [211]
Succinic acid	P	Pagel et al. [223, 224]
	A	Manenok et al. [212]
	A	Vieux et al. [213, 214]
maleic acid	P	Pagel et al. [223, 224]
	A	Pyatnikskii et al. [211]
malic acid	P	Pagel et al. [223, 224]
tartaric acid	P	Pagel et al. [223, 224]
	A	Pyatnitskii et al. [211, 221, 222]
citric acid	P	Pagel et al. [223, 224]
	A	Pyatnitskii eta l. [211, 219–222]
	P	Wennersten [204]
	A	Wennersten [205]
	P, A	Jian Yu-Ming et al. [206]
	A	Friesen et al. [207]

A review on propionic, lactic, pyruvic, succinic, fumaric, malic, itaconic, tartaric, isocitric and citric acids is given by Kertes and King ([9] in Chapter 2) and succinic acid by Tamada et al. [225]

Table 4.3. Dissociation constants pK of carboxylic acids at 25 °C ([9] in Chapter 2)

formic acid*	3.75	maleic acid	1.93, 6.14
acetic acid	4.75	malic acid	3.22, 4.70
propionic acid	4.87	itaconic acid	3.65, 5.13
lactic acid	3.86	tartaric acid	3.01, 4.38
pyruvic acid	2.49	citric acid	3.13, 4.76, 6.40
succinic acid	4.20, 5.64	isocitric acid	3.29, 4.71, 640
fumaric acid	3.02, 4.38		

* at 20 °C

K_{di} in the dilute concentration range, where Eq. (2.25) in Chapter 2 holds true, are compiled in Table 4.4. One can observe that the conventional extraction systems using water-immiscible solvents are relatively inefficient for acid recovery from dilute solutions.

Table 4.4. Partition coefficients P_c and dimerization constants K_{di} of carboxylic acids at 25 °C ([9] of Chapter 2)

Acid/solvent Propionic acid in	P_c		K_{di} $(l\,mol^{-1})$
n-Hexane	0.005		9000
Cyclohexane	0.006		6500
Benzene	0.043		190
Toluene	0.034		230
Xylene	0.030		310
Carbon tetrachloride	0.015		940
Chloroform	0.11		30
Nitrobenzene	0.16		11
Diethyl ether	1.75		0.1
Diisoproyl ether	0.80		0.5
Methylisobutyl ketone	2.15		
Cyclohexanone	3.30		
n-Butanol	3.20		
n-Pentanol	2.95		
Lactic acid in Diethyl ether	0.10	Isobutanol	0.66
Diisopropyl ether	0.04	n-Pentanol	0.40
Methylisobutyl ketone	0.14	n-Hexanol	0.37
n-Octanol	0.32		
Succinic acid in Diethyl ether	0.15	Isobutanol	0.96
Methylisobutyl ketone	0.19	n-Pentanol	0.66
n-Butanol	1.20	n-Octanol	0.26
Fumaric acid in Diethyl ether	1.50	n-Butanol	3.30
Methylisobutyl ketone	1.40	Isobutanol	4.60
Maleic acid in Diethyl ether	0.15	Isobutanol	0.92
Methylisobutyl ketone	0.21		
Malic acid in Diethyl ether	0.02	Isobutanol	0.36
Methylisobutyl ketone	0.04		
Itaconic acid in Diethyl ether	0.35	Isobutanol	1.80
Methylisobutyl ketone	0.55		
Tartaric acid in Diethyl ether	0.03	n-Butanol	0.16
Methylisobutyl ketone	0.02		
Citric acid in Diethyl ether	0.009	n-Butanol	0.29
Methylisobutyl ketone	0.09	Isobutanol	0.30

The data of acetic acid are given in Table 4.5

Not only the distribution coefficients, but also the mutual solubilities of the phases, the dissociation of acids and the chemical nature of extractants, especially the donor-acceptor capability, as well as the thermodynamic functions of transfer are discussed in this review [9] (in Chapter 2).

First of all, the acetic acid recovery is dealt with. The conventional extraction technology uses a combination of extraction and azeotropic distillation at low acetic acid concentrations and azeotropic distillation at high concentrations [187–190]. According to Eaglesfield [189] Brockhaus and Förster [188] the upper limit of the extraction is at 35 wt%. At low acid concentrations, the extraction step serves as preconcentrator of the feed to the azeotropic distillation column. One usually chooses a solvent for extraction which itself serves as entrainer for the azeotropic distillation. For low-boiling solvents, the bottom product is the acetic acid with 0.1 to 0.5 wt% solvent residue, and the head product is the solvent-water mixture, which is separated in a phase separator. To recover the dissolved solvent from the aqueous phase, the water and the solvent are separated in a solvent recovery distillation column [187–190].

Two important aspects have to be considered for the solvent selection: the equilibrium distribution coefficient K_D of acetic acid in the solvent/aqueous phase system and the boiling point of the solvent with regard to that of the acetic acid. Treybal [191] compiled K_D (weight fraction acid in the solvent phase/weight fraction acid in the aqueous phase) values for different solvents for high dilution. In Table 4.5 the K_D ranges are given for different solvents. The equilibrium distribution coefficient undergoes a continual transition between the extreme values shown. The lower molecular-weight solvent shows the highest value of K_D. The highest K_D values are obtained by alcohols, but, they react with the acetic acid and therefore are not used. The next highest K_D values are obtained by ketones. However, they do not form good azeotropic mixtures for water removal. Therefore, esters are the most commonly used solvents [192]. The presence of salts in the aqueous phase can increase K_D considerably [196]. Because of the low K_D values relatively high solvent-to-feed ratios (1.6 to 4.0) are used. Since the solvent is also an entrainer for the azeotropic distillation, it is important that the solvent has a large capacity for water in azeotropic distillation and a low solubility in water to reduce the load on the solvent recovery stripper [192].

One can choose a solvent with a boiling point lower or higher than that of acetic acid. The pros and cons for low and high-boiling-point solvents were specially considered [190].

Lactic acid plays an important role in the food industry. It is recovered from the cultivation by means of its low-solubilty Ca-salt. Ca-lactate is treated with sulfuric acid to obtain the free acid. $CaSO_4$ precipitate has to be removed. The filtrate, an aqueous solution of lactic acid, is concentrated by vacuum evaporation. Volatile byproducts (acetic and propionic acids) are removed in this way.

Table 4.5. K_D values of acetic acid in different solvents [192]

Solvents	K_D range
n-Alcohols (C_4–C_8)	1.68–0.64
Ketones (C_4–C_{10})	1.20–0.61
Acetates (C_4–C_{10})	0.89–0.17
Ethers (C_4–C_8)	0.63–0.14

The data of acetic acid are given in Table 4.5

The concentrated solution is decolorized on chark coal and deionized on ion exchanger columns. This treatment yields a concentrated (80 wt%) lactic acid. On account of its highly hydrophilic character, its purification with solvent extraction by carbon bonded oxygen donor extractants is not practical.

Citric acid is the most widely used organic acid in the food and pharmaceutical industries. Its alkaline salts are widely used as buffers. Esters of citric acid are used as plasticisers for polymers.

Solvent extraction can be an alternative for the recovery of citric acid from the fermentation broth as a Ca-salt precipitate, from which the free acid is liberated by sulfuric acid. However, the distribution coefficients of the carbon-bounded oxygen donor solvents are very low (Table 4.6). Moreover, these solvents are relatively soluble in water and often form azeotropes with water. Therefore, citric acid extraction with these solvents is not economical.

Phosphorus-bonded, oxygen-donor extractants contain a phosphoryl group which is a stronger Lewis base than the carbon-bonded oxygen group. Furthermore, extractants belonging to this group coextract less water and are less soluble in water than the extractants of the carbon-bonded oxygen group. E.g., in the equilibrium system of 30 wt% TOPO in Chevron 25 and 10% aqueous acetic acid, the TOPO concentration in the aqueous phase is less than 1 ppm. The solubility of water in a 45 wt% TOPO, 55 wt% 2-heptane system depends on the acetic acid concentration. It increases from about 0.25 wt% (at 1 wt% acetic acid) to about 2 wt% (at about 10 wt% acetic acid) ([7] in Chapter 2).

The basicity of the phosphoryl oxygen increases in the order: trialkyl phosphate $((RO)_3 P = O)$, dialkyl phosphonate $((RO)_2 R P = O)$, alkyl dialkyl phosphinate $((RO)R_2 P = O)$, trialkyl phosphine oxide $(R_3 P = O)$. According to Wardell and King ([6], in Chapter 2) the basicity of the extractants increases with the number of butoxy groups $(-O CH_2 . CH_2 . CH_2 . CH_3)$ for the extraction of acetic acid (Table 4.7).

The Mass Action Law Equilibria for phosphorus-bonded oxygen-donor reactants are also given in Chapter 2.1.1 (Eqs. (2.27), (2.28) (2.29), and (2.34)–(2.37) in Chapter 2).

Weak acids are extracted by organophosporus compounds with significantly higher distribution ratios than by carbon-bonded oxygen donor extractants under comparable experimental conditions. Several authors investigated the

Table 4.6. Distribution coefficients of citric acid in different solvents [208]

Solvents	K_D
n-Butanol	0.29
Ethylacetate	0.1
Ethylether	0.1
Methyl-isobutyl ketone	0.1
Methyl-ethyl ketone	0.33
Cyclohexanone	0.21

Table 4.7. Effect of the number of butoxy groups on the distribution of acetic acid from 0.5 wt% water solution ([6] in Chapter 2)

Extractant	Number of butoxy groups in extractant	Diluent	K_D
Tributylphosphate	3	—	2.3
Dibutylphosphate	2	—	2.7
Butyl-dibutyl phosphinate	1	—	—
Tributylphosphine oxide	0	37.3 wt% Chevron 25	4.4

equilibrium constants and distribution coefficients between aliphatic carboxylic acids and organophosphorus compounds ([6–9] in Chapter 2) and [192, 194, 195, 200].

In Table 4.8, some K_D values for acetic acid with tributyl phosphate (TBP) and trioctyl phosphine oxide (TOPO) as extractants with different diluents at a high dilution are compared ([6, 7] in Chapter 2). K_D values are higher than unity, and they depend on the diluents. Furthermore, the K_D value is considerably influenced by the acid concentration (Table 4.9). With increasing acid concentration, the K_D value is strongly reduced ([7] in Chapter 2). Therefore, the main use of these extractants lies in the removal of acid traces from aqueuos solutions (e.g., in wastewater treatment) [195].

Lactic acid can be extracted by a mixture of tri- and di-hexyl and tri- and di-octyl phosphinoxides (Cyanex 923 Cyanamid) as extractants in kerosene.A one to one acid-carrier complex is formed [238]. However, for lactic acid recovery mainly amine carriers were applied.

Few carboxylic acid systems have been investigated extensively enough to permit the evaluation of the association constant K_S (using Eq. (2.35) in Chapter 2) (Table 4.10).

Table 4.8. K_D values of acetic acid with Trioctylphosphine oxide (TOPO) in different diluents at high dilution ([6, 7] of Chapter 2)

Extractant	Diluent	K_D
50% TOPO	2-ethyl-1-hexanol	1.12
50% TOPO	2-heptanone	2.83
50% TOPO	Chevron 25	2.01

Table 4.9. Influence of the acetic acid concentration on the K_D value with 22 wt% TOPO in Chevron 25 ([7] of Chapter 2)

Acid concentration (wt%)	K_D
0.189	3.12
1.27	1.33
3.20	0.766
7.45	0.450

Table 4.10. Equilibrium constants K_e for the formation of solvates between carboxylic acid and organophosphorus extractants ([9] of Chapter 2)

Acid	Extractant	Diluent	Solvation number	K_e	$T(°C)$
Propionic	TOPO	n-Hexane	1	59.0	25
Succinic	TOPO	n-Hexane	2	152.0	25
Malic	TBP	n-Dodecane	2	0.11	20
Tartaric	TBP	n-Dodecane	2	0.04	20
Lactic	TBP	n-Dodecane	–	0.26	20
Lactic	DEHPA	Diethylbenzene	1	38.0	–
Citric	TBP	Carbon tetrachloride	3	0.04	25

The extraction power increases considerably as the number of direct C-P-linkages increases. Thus the acid adduct with TOPO is two orders of magnitude more stable than that with TBP ([9] in Chapter 2). The solvation numbers of the acid in Table 4.10 seem to be equal to the number of carboxyl groups on the acid molecule. This indicates a stoichiometric association between an individual phosphoryl group and an individual carboxylic group.

The distribution ratio D_c decreases initially rather sharply with an increasing acid concentration ([9] in Chapter 2). The effect of the diluent on the extractive properties of the alkylphosphates is much less than expected (Table 4.11). This table shows that for the extraction of citric acid from 0.26 mol l^{-1} aqueous solution by an organic phase containing 50 vol% TBP, its distribution ratio is 0.56 with diisopropylether as diluent, 0.5 with methyl isobutylketone, 0.33 with SHELLSOL H and 0.31 with SHELLSOL A (SHELLSOL H is aliphatic and SHELLSOL A is an aromatic hydrocarbon mixture) [204].

Furthermore, these measurements indicate a strong influence of the temperature; with increasing temperature K_D diminishes. The influence of the solute concentration is relatively slight, especially at high temperatures.

Table 4.11. Distribution coefficients K_D for the extraction of citric acid from water solution to an organic solvent with different extractants and diluents [204]

Extractant	Diluent	Acid conc. (M)	K_D	$T(°C)$
0.4 M TOPO	50 vol% n-Heptane 50 vol% n-Octanol	0.21	0.01	20.6
0.2 M TOPO	Shellsol H	0.072	0.28	21.0
50 vol% TBP	50 vol% Diisopropylether	0.26	0.56	17.5
		0.26	0.23	60.0
50 vol% TBP	50 vol% MIBK	0.26	0.50	17.5
		0.26	0.20	60.0
50 vol% TBP	50 vol% Shellsol H	0.26	0.33	17.5
		0.26	0.10	60.0
50 vol% TBP	50 vol% Shellsol A	0.26	0.31	17.5
		0.26	0.09	60.0

When using particular diluents, two organic phases are formed (e.g., in the case of lactic acid). (This phenomenon is discussed in connection with the use of aliphatic amine extractants). Only two groups have carried out measurements to recover citric acid from fermentation broths [204, 206].

Both of them complain about the formation of stable emulsions. When using a small laboratory mixer-settler, the settler was flooded in spite of very low throughputs [204].

Tests were also carried out in a Karr-column (height 3 m, 5 cm in diameter) at 20 °C. In the continuous aqueous phase, (with 37 g l^{-1} citric acid in feed) the organic phase was dispersed. If the phase ratio V_o (volume of the organic phase) to V_a (volume of the aqueous phase) exceeded 1.4, phase inversion occurred, and the column was flooded. It was not possible to run the column with the dispersed aqueous phase. For phase ratios above 1.4, a three-stage mixer-settler battery was used, in which ALFA LAVAL centrifuges were used as settlers. With 70% TBP in SHELLSOL H, very poor results were attained [204].

Lately citric acid was also purified by electrodialysis [227[. Lactic acid is also recovered from the fermentation broth by means of its low soluble Ca-salt. Ca-lactate is treated with sulphuric acid to obtain the free acid. CaSO$_4$ has to be removed similar to citric acid production. To avoid the formation of CaSO$_4$ several research groups performed research for lactic and citric acid recovery.

Lately electrodialysis with bipolar membrane has been used to obtain the free lactic and citric acid, respectively [228–232]. On extraction of lactic acid by liquid surfactant membrane and tri-n-octylphosphineoxide (TOPO) and tri-n-butylphosphate was reported by Hano et al. [233].

Smith and Page [209] reported on acid-binding properties of *long-chain aliphatic amines* which are based on their insolubility in aqueous solutions and solubility in organic solvents. When using these high molecular weight aliphatic amines, the extraction occurs by proton transfer or by ion-pair formation. The Mass Action Law Equilibria are given in Chapter 2.1.1 (Eqs. (2.39)–(2.92) in Chapter 2).

The acid extracted into the amine-containing organic phase is regarded as ammonium salt. Therefore, the extent of ion-pair association between the alkylammonium cation and acid anion is the measure of extractability or stability of the organic phase species ([9] in Chapter 2). The extent to which the organic phase (amine + diluent) can be loaded with acid is expressed as the loading ratio Z:

$$z = \frac{C_{HSo}}{c_{Ao}} \tag{4.1}$$

where C_{Ao} is the amine concentration and C_{HSo} is the solute concentration in the organic phase. z should be corrected for the acid extracted into the diluent alone, if a diluent is used. The value of Z depends on the strength of the acid-base interaction and on the aqueous acid concentration and is independent of the amine concentration in an inert diluent, as long as it exists in excess. The relationship between the loading ratio and the equilibrium constant K_e of

Eq. (2.40) in Chapter 2 is given by

$$\frac{z}{1-z} = K_e[HS]_a \tag{4.2}$$

at low association concentrations in the organic phase. With linear plot of $z/(1-z)$ against $[HA]_a K_e$ the slope yields K_e in this case.

At higher amine concentrations in inert diluents, a molecular association of the alkylammonium salts AHS is formed in the organic phase:

$$q\,AHS \overset{K_q}{\rightleftharpoons} (AHS)_q, \quad K_q = \frac{[(AHS)_q]}{[AHS]^q} \tag{4.3}$$

q depends on the chemical nature of the salt, its concentration, the nature of diluent and the temperature ([9] in Chapter 2).

Ricker et al. ([7] in Chapter 2) investigated the extraction of acetic acid with different high-molecular-weight primary, secondary and tertiary amines (Table 4.12). Their investigations indicate that primary amines have a high mutual solubility with water, secondary amines yield high values of K_D (Table 4.13), actually they provide the highest K_D value ($K_D = 160$ with Adogen 283-D in

Table 4.12. Extractants used for acetic acid extraction by Ricker, Michaels and King ([7] of Chapter 2)

Extractant	Manufacturer	Type	Chemical composition
Primine JMT	Rohm & Haas	prim. Amine	highly branched, 18–24 C atoms
Amberlite LA-3	Rohm & Haas	prim. Amine	branched chain ave. MW = 353
Adogen 283-D	Ashland Chem.	Sec. Amine	di-tridecyl amine branched
Amberlite LA-2	Rohm & Haas	sec. Amine	highly branched, 12–15 C atoms
Amberlite LA-1	Rohm & Haas	sec. Amine	highly branched,
Adogen 383	Ashland Chem.	tert. Amine	tri-lauryl, straight chain
Adogen 363	Ashland Chem.	tert. Amine	tri C_8–C_{10} isomers, straight chain
Adogen 367-D	Ashland Chem.	tert. Amine	di-methyl coco amine
Adogen 368	Ashland Chem.	tert. Amine	tri C_8, C_{10}, C_{12} isomers, straight chain
Adogen 345-D	Ashland Chem.	tert. Amine	di-methyl hydrogenated tallow amine
Adogen 381	Ashland Chem.	tert. Amine	tri-isooctyl amine
Methylated Adogen 283-D	prepared	tert. Amine	methyl di-tridecyl amine
Alamine 336	General Mills	tert. Amine	tri C_8 and C_{10} straight chains
TOPO	American Cyanamid	Phosphine oxide	tri-n-octyl

Table 4.13. Extraction of acetic acid by several secondary amines ([7] of Chapter 2)

Amine	Vol % of amine	Diluent	Equil. acid conc. (wt. %)	K_D
Amberlite LA-1	50	Chevron25	1.16	1.27
			3.78	2.26
	30	chloroform	0.0469	4.48
	100	none	2.52	4.22
Amberlite LA-2	50	Chevron 25	0.53	3.82
			2.19	4.48
	30	chloroform	0.0218	9.86
	100	none	1.83	6.49
Adogen 283-D	50	Chevron 25	0.983	9.65
			2.63	4.55
			3.85	3.46
	30	chloroform	0.0133	32.11
	100	none	0.383	33.4

2-ethyl-1-hexanol) ([7] in Chapter 2); however, they are subject to amide formation during regeneration by distillation. Tertiary amines are very effective (Table 4.14), but generally give lower K_D values than the secondary amines.

As the extractants are fairly expensive, it is important that their loss, due to their solubility in the aqueous phase, is low. Long-chain tertiary amines are practically insoluble in the aqueous phase at an acid concentration below 10 wt%. E.g., in a 1 wt% solution, the solubility of Alamine 336 is less than 10 ppm ([17] in Chapter 2). At a constant amine concentration in the diluent, K_D diminishes with increasing acid concentration. On the other hand, the increase of amine concentration at a constant acid concentration improves K_D at first, but above a certain level, which depends on the diluent, the amine

Table 4.14. Extraction of acetic acid by several tertiary amines ([7] of Chapter 2)

Amine	Vol % of amine	Diluent	Equil. acid conc. (wt. %)	K_D
Adogen 381	20	2-heptanone	3.23	1.72
	30	chloroform	0.0448	7.79
Adogen 364	20	2-heptanone	2.90	1.98
	30	chloroform	0.0381	9.69
Adogen 368	20	2-heptanone	3.11	1.83
	30	chloroform	0.0447	8.28
Adogen 363	20	2-heptanone	3.44	1.58
	30	chloroform	0.0623	7.82
Alamine 336	20	2-heptanone	2.99	2.24
	30	chloroform	0.0465	9.68
methyl di-tridecyl	50	methyl iso-amyl ketone	2.91	3.54

and acid concentrations, K_D decreases, which is probably due to the increasing nonpolarity of the organic phase.

Diluents reduce the viscosity of the organic phase and can be used to control the density difference between the phases and the interfacial tension. As the diluents are relatively poor solvents for the acid-base complexes, the use of a second diluent (so-called modifier) can improve the solvation of the complex. Hence, K_D is higher for intermediate amine concentrations in the diluent mixture than in the pure diluents (synergism). The effectiveness of diluents tends to increase with increasing polarity ([6] in Chapter 2).

Alcohol diluents yield the highest K_D values, but are subject to esterification with the acid. In general, the K_D value diminishes in the following order: alcohols, ketones, esters, hydrocarbons, if they are used as diluents. The water content of the organic phase depends on the acetic acid concentration, e.g., the solubility of water increases from about 0.25 wt% (with 1% acid) to about 2 wt% (with 10% acid) in the organic phase (20 to 65 vol% Alamine 336 in DIBK) ([7] in Chapter 2).

There is an optimal molecular weight of the extractant which is a compromise between a high K_D (low molecular weight) and a low solubility in the aqueous phase (high molecular weight). Alamine 336 appears to be close to this optimum molecular weight. For the diluent, one has to compromise between a high K_D (low-molecular weight) and a low solubility in water or a sufficiently low volatility relative to acetic acid for solvent regeneration by distillation (high molecular weight). Among the ketones, diisobutyl ketone (DIBK) seems to be the optimum, as the relative volatility of acetic acid to DIBK is about 3.5 for an organic phase, 40 vol% Alamine 336 in DIBK. With this system $K_D = 2.5$ was attained with 1 wt% acetic acid in the aqueous phase and 50 vol% Alamine 336 in DIBK as organic phase ([8] in Chapter 2).

On the basis of costs in 1978, Ricker et al. ([8] in Chapter 2) estimated total operating costs as US$1.9/m^3 water for extraction of acetic acid from 5 wt% aqueous solution into solvent mixtures of 50% Alamine 336 in DIBK and 40 wt% TOPO in 2-heptanone, respectively. This would lower the economically recoverable feed concentration by a factor of two in comparison with use of ethylacetate as solvent.

As already mentioned, the use of polar diluents for acetic acid extraction increases the K_D values considerably: e.g., with 2-ethylhexanol, the K_D value is improved by a factor of 23 and with chloroform by a factor of 11 ([6, 7] in Chapter 2). Furthermore, the addition of isodecanol to amine extractants: tri-n-decyl/n-octylamine (Hostarex A 327, Hoechst) and/or tri-isooctylamine (Hostarex A 321, Hoechst) in a hydrocarbon diluent (SHELLSOL T) improved the K_D value at low acetic acid concentration considerably [193]. Wojtech and Mayer [195] investigated the influence of several alcohols and phenols as diluents for a 50:50 mixture of tri-n-octyl/tri-n-decylamine (Hostarex A 327, Hoechst, and they found the highest distribution coefficient with 90% p- and 10% o-nonylphenol and in the following order: n-butanol > n-pentanol > 3-methyl butanol (isoamylalcohol) > n-octanol > 2-ethylhexanol > isodecanol with decreasing effect. The advantage of the acetic nonylphenol to alcohols is that it does not form an ester with acetic acid during distillation.

A sharp maximum of the distribution coefficient ($K_D = 300$) appeared as a function of the extractant content at 30 wt% Hostarex A 327 and 70 wt% nonylphenol (mol ratio 4:1) with 1.4 wt% acetic acid and a phase ratio $V_a/V_o = 2$ to 1 at 25 °C. An alternate system with a lower distribution coefficient, but with a higher loading capacity was worked out: 60 wt% n-pentanol and/or 3-methylbutanol and 40 wt% Hostarex A 327 (mol ratio 7:1) with a distribution coefficient of 30 at 1.4 wt% acetic acid. The latter system was used for the recovery of acetic acid from waste water because of the higher loading capacity. In a single-stage extractor, a 99% degree of extraction was attained. Since both the extractant and the diluent have boiling points above 300°C, the distillative separation of acetic acid (b.p. 118 °C) is easy.

With this extractant/diluent combination, the distribution ratios of other acids (butyric, malonic, lactic and glycolic acids) were also improved by a factors of ten to 100 [195].

Siebenhofer and Marr [194] combined the two types of extractants: an organophosphorus compound (dioctyl phosphonate) and a high molecular weight amine (trioctyl amine) in a hydrocarbon diluent (SHELLSOL T) for the extraction of acetic acid. The replacement of the organic phase (20 vol% triisooctylamine and 80 vol% SHELLSOL T) by a new one (20 vol% triisooctylamine, 40 vol% dioctyloctyl phosphonate and 40 vol% SHELLSOL T) improved the distribution ratio by a factor as high as ten.

As in the case, the solvation effect was combined with the ion-pair formation, the stoichiometry of the extraction was influenced by the combination of these two types of extractants, as expected. Lactic acid was extracted by different long chain amines [238]. The highest performance was obtained with Hostarex A 327 (50:50 mixture of n-trioctyl- and n-tridecyl amines) (Hoechst) and Hoe F 2562 (di-tridecylamine) (Hoechst) and isodecanol as modifier as extractants in kerosene.

The lactic acid (HS) is distributed between the aqueous (a) and organic (o) phases, dissociates to anion L^- and proton H_3O^- and reacts with the carrier (A):

1) Physical extraction:

$$HS_{a,p} \overset{K_p}{\rightleftharpoons} HS_{o,p} \tag{4.4}$$

$$k_p = [HS]_{o,p} V_o / [HS]_{a,p} V_a \tag{4.4a}$$

2) Dissociation

$$HS_a + H_2O \overset{k_a}{\rightleftharpoons} H_3O + S_a^- \tag{4.5}$$

$$[HS]_a = \frac{c_o[HS]}{1 + 10(pH - pK_a)} \tag{4.5a}$$

3) Complex formation ($V_a : V_o = 1 : 1$)

$$HS_a + (A)_o \overset{k_1}{\rightleftharpoons} (HSA)_o \tag{4.6}$$

$$k_1 = \frac{[HSA]_o}{[HS]_a [A]_o} \tag{4.6a}$$

$$HS_a + (HSA)_o \overset{k_2}{\rightleftharpoons} ((HS)_2 A)_o \tag{4.7}$$

$$k_2 = \frac{[(HS)_2 A]_o}{[HS]_a^2 [HSA]_o} \tag{4.7a}$$

where k_1 and k_2 are the complex formation constants with one and two lactic acids [238].

With Hostarex A 327 and 10–40 wt% of lactic acid the following constants were determined:

$k_1 = 5.59$ [$l\,mol^{-1}$] (5 wt% isodecanol), 11.96 (10 wt% isodecanol), 63.2 (40 wt% isodecanol) and $k_2 = 3.05$ [$l\,mol^{-1}$] (5 wt% isodecanol), 2.85 (10 wt% isodecanol), 7.33 (40 wt% isodecanol), as well as with HOE F 2562 and 5 wt% isodecanol: $k_1 = 49.1$ and $k_2 = 13.8$ [238].

Frieling [238] developed relationships for the calculation of k_1 and k_2 for these systems. There is a good agreement between the calculated and measured complex formation constants. In extractant-butylacetate diluent, the partition coefficients are mainly controlled by the partition coefficient of lactic acid between the aqueous and butyl acetate phases. The partition coefficient $k_p = 0.28$ at 23 °C differs from the coefficient: $k_p = 0.11$ at 20 °C evaluated by Holten [239]. For the Hostarex A 327/butylacetate system, the complex formation constants are: $k_1 = 8.82\,l\,mol^{-1}$ and $k_2 = 2.83\,l\,mol^{-1}$ [238].

Wennersten [205] investigated the citric acid extraction with three tertiary amines, i.e., tribenzylamine, tri-dodecyl amine and Alamine 336 (tri-*n*-octyl-amine) in different diluents. The influence of diluent, amine structure and acid concentrations on the distribution coefficient K_D were investigated. In Tables 4.14, 4.15, the effect of the diluents on the K_D of the citric acid with Alamine 336 and tridodecylamine as extractants are shown at different temperatures and acid concentrations. At low acid concentrations, a strong diluent effect can be recognized. The K_D value is a function of the polarity of the diluent and its ability to form hydrogen bonds. At higher acid concentrations, however, this diluent effect diminishes considerably. For highly complexed amines, the K_D value depends only slightly on the acid concentration.

The K_D value is considerably reduced with increasing temperature. This behaviour of K_D is used to reextract the citric acid into the aqueous phase by temperature increase. Tribenzylamine yields a very low distribution coefficient, therefore, it is not dealt with here.

In Tables 4.16, 4.17, the distribution coefficients are shown for ISOPAR H (paraffine/kerosene mixture), MIBK and *n*-butyl chloride as diluents and Alamine 336 as extractant at different citric acid concentrations. For each set of

Table 4.15. Effect of diluent on the distribution coefficient of citric acid with 50 vol % Alamine 336 [205]

Diluent	Acid conc. (M)	K_D	$T(°C)$
n-hexane	$14.3 \ 10^{-3}$	2.9	25
	0.382	1.9	25
	$37.9 \ 10^{-3}$	0.3	60
toluene	$4.93 \ 10^{-3}$	10.8	25
	0.358	2.1	25
	$22.1 \ 10^{-3}$	1.4	60
1,1,1-tri-chlorethane	$3.37 \ 10^{-3}$	16.4	25
	0.348	2.2	25
	$12.7 \ 10^{-3}$	3.5	60
ethyl acetate	$2.45 \ 10^{-3}$	23.0	25
	0.275	3.1	25
	$15.7 \ 10^{-3}$	2.5	60
MIBK	$2.07 \ 10^{-3}$	27.5	25
	0.276	3.1	25
	$5.17 \ 10^{-3}$	11.5	60
cyclo-hexanone	$1.87 \ 10^{-3}$	30.5	25
	0.220	4.2	25
	$3.2 \ 10^{-3}$	17.3	60
2-ethyl hexanol	$0.83 \ 10^{-3}$	70.0	25
	0.351	2.1	25
	$1.97 \ 10^{-3}$	28.9	60
iso-amyl alcohol	$0.67 \ 10^{-3}$	87.8	25
	0.305	2.7	25
	$1.37 \ 10^{-3}$	42.2	60
n-butanol	$0.35 \ 10^{-3}$	168.3	25
	0.274	3.1	25
	$0.8 \ 10^{-3}$	72.9	60

Table 4.16. Effect of diluent on the distribution coefficient of citric acid with tri-dodecylamine [205]

Diluent	Acid conc. (M 10^3)	K_D	$T(°C)$
MIBK	1.7	34.0	25
	9.87	4.8	60
2-Ethylhexanol	0.33	177	25
	1.8	31.7	60
Toluene	8.20	6.0	25
	329	0.55	60
Chloroform	0.43	136	25
	3.13	17.7	60

data, a maximum of the distribution coefficient was found at a citric acid concentration at which the mol ratio of citric acid to Alamine 336 is about 0.2 to 0.4. This corresponds to one mol amine per one mol carboxyl group. Rückl et al. [226] found an optimum with the blend: 30% Hostarex A 324, 30% isodecanol and 40% alcanes.

Table 4.17. Effect of diluent and acid concentration on the distribution coefficient of citric acid with Alamine 336 at 25 °C [205]

Extractant/Diluent (vol/vol)	Equil. acid conc. (mM) in aqu. phase	K_D
Alamine 336	28	9.6
Isopar H 1:1	67	7.99
	128	4.75
	193	3.48
	295	2.34
	390	1.85
	621	1.19
	1000	0.76
Alamine 336	7.7	38.4
Isopar H	22.7	26.3
MIBK 2:1:1	62	11.1
	128	5.9
	230	3.36
	323	2.50
	565	1.44
	954	0.86
Alamine 336	8.7	33.8
n-butylchloride 1:1	27.3	21.6
	75	8.97
	151	4.81
	258	2.86
	357	2.13
	597	1.29
	1001	0.76

According to Wennersten, at citric acid concentrations above the mol ratio one amine to one carboxyl group and without diluents or with hydrocarbon diluents, a second organic phase can appear (Table 4.18). The upper organic layer consists mainly of an amine and the lower one of an acid-amine complex. This phase separation is due to the low solubility of the acid-amine complex in the amine extractant. The lower organic phase was extremely viscous and the acid could not be reextracted with water, even at a high temperature. Therefore, the formation of a second organic phase must be avoided by using a high ratio of organic-to-aqueous phases in the extractor.

After the extraction of the citric acid from the fermentation broth, the color of the organic phase and the phase separation rates depend on the diluents. Polar diluents yield colored organic phases and moderate or low separation rates. Hydrocarbon diluents (n-hexane) do not extract colored compounds from the broth and exhibit fast phase separation. Therefore, they have a high "selectivity" with regard to the coextraction of accompanying broth components (Table 4.19). Wennersten [205] recommends hexane as a diluent because of its low toxicity, low solubility in water, low viscosity and density as well as high stability, and in spite of a low distribution coefficient of the citric acid in hydrocarbon diluents.

Table 4.18. Effect of diluent and acid concentration on the distribution coefficient of citric acid with Alamine 336 at 60 °C [205]

Extractant/Diluent (volume/volume)	Equil. acid conc. (mM) in the aqu. phase	K_D
Alamine 336	201	1.86
Isopar H 1:	272	1.57
	331	1.51
	415	1.30
	497	1.18
	700	0.92
	1067	0.63
Alamine 336	28	5.22
Isopar H	79.3	6.63
MIBK 2:1:1	186	3.65
	369	2.03
	600	1.28
	982	0.8
Alamine 336	97.3	5.42
n-butyl chloride 1:1	146	4.0
	213	3.04
	312	2.14
	405	1.74
	636	1.13
	1027	0.71

Table 4.19. Influence of the diluent amount and composition on the formation of the second organic phase with 50% Alamine 336 in diluent and 0.83 M citric acid in the initial aqueous phase [205]

Diluent	Number of phases
no diluent	3
25 ml Isopar	3
20 ml Isopar + 5 ml toluene	3
15 ml Isopar + 10 ml toluene	3
10 ml Isopar + 15 ml toluene	2
5 ml Isopar + 20 ml toluene	2
20 ml Isopar + 5 ml MIBK	3
15 ml Isopar + 10 ml MIBK	2
25 ml MIBK	2
25 ml Shellsol A	2

In a 3 m height and 0.05 m in diameter Karr column, at a stroke frequency of 60 min^{-1} and stroke amplitude of 0.012 m, the citric acid extraction from fermentation broth was performed at 25 °C with a 30 vol% Alamine 336 and a 70 vol% SHELLSOL A (a kerosine with high aromatic content). With 0.58 M citric acid in feed, a feed rate of 1.8 l h^{-1}, and a flow rate of the organic phase of 3.6 l h^{-1}, a 97% degree of extraction was achieved [205]. The extraction was carried out with the pellet-containing fermentation broth. After starting difficulties, a steady-state operation could be maintained.

The reextraction of the citric acid from the organic phase was carried out in an ALFA LAVAL plate heat exchanger, which was used as static mixer, with solvent phase flow rates of $1.2 \, l \, min^{-1}$ and stripping phase flow rates of $0.4 \, l \, min^{-1}$ at 63 °C. No degree of reextraction was given [205].

Jain Yu-Ming et al. [206] also investigated the extraction of citric. acid from fermentation broths by different extractants (Table 4.20). They also found that amines have the highest distribution coefficients, but they state that amines have the following drawbacks:

– because of the very high distribution coefficient, the reextraction is difficult,
– they are toxic, therefore they cannot be recommended for the production of citric acid for edible purpose,
– they form stable emulsions.

The emulsion formation can be reduced by removing the main part of the dissolved proteins from the broth by adsorption on active carbon. Trialkyl phosphine oxide was used as extractant for the recovery of citric acid from fermentation broths, mainly because of its nontoxicity (according to the British Pharmaceutical Codex (1977)) and its odorless properties.

The recovery of citric acid begins with the separation from the mold by filtration, removal of the proteins by adsorption on active carbon, before the extraction is performed. Reextraction is carried out with a water and alkaline solution. The extract is purified on an ion exchanger and crystallized [206].

Citric acid extraction with trialkylamine and MIBK was theoretically invest-igated and mathematically modeled by Bizek et al. [234]. They found that the extraction mechanisms are mixed: non-dissociated as well as dissociated acid forms an amine complex. Bizek et al. [234] reported on the theoretical investiga-tion and mathematical modelling of citric acid extraction with trialkylamine in MIBK. They found that non-dissociated acid as well as dissociated acid forms an amine complex and participates in the extraction.

Table 4.20. Distribution coefficient of citric acid extracted from filtered fermentation broth with 100 g/l initial concentration with different extractants at phase ratio of 1 : 1 and at 25 °C [206]

Extractants	K_D
iso-amyl alcohol	0.13
cyclohexanone	0.37
MIBK	0.20
tertiary amine (C_8–C_{10})	50.0
trioctylamine	15.4
TBP	1.49
tryalkylphosphine oxide	7.61
dialkylphosphonate	0.82
dibutylsulfoxide	3.24

Emulsion liquid membranes with Alamin 336 (trioctylamine) was used by Boey et al. [235] and supported liquid membranes with Alamin 336 by Sirman et al. [236] to recover citric acid from fermentation broth.

A supported liquid membrane technique was recommended by Friesen et al. [207] for the recovery of citric acid from fermentation broths too. Tertiary amines were used as extractants, long-chain alcohols as modifiers and hydrocarbons as diluents. This blend was immobilised within the pores of microporous membranes. The citric acid was complexed on one side of the membrane and reextracted on the other side. High purity citric acid was recovered in a single-step operation from the broth.

Citric acid extraction with microporous hydrophobic hollow fibers was performed by Basu and Sirkar [237]. The aqueous citric acid solution was on one side and the organic phase with trioctylamine in MIBK as extractant on the other side of the membrane (Celgard X-20, 240 μm id, 290 μm od, Celanese). However, in laboratory scale equipments the recovery process has a very low rate. According to the authors in large scale equipments high recovery rates can be obtained.

In lactic acid production with several *lactobacilli* the supplementation of the medium with acetic and citric acids increases the productivity [240]. Therefore the separation of acetic, lactic and citric acids could be of importance. In Figs. 4.4, 4.5, the degree of coextractions of acetic, lactic and citric acids with 30 wt% Hostarex A 327 and 10 wt% isodecanol and 30 wt% Cyanex 923 at different temperatures are shown. In is not possible to separate these three acids with Hostarex A 327/isodecanol in single stage. The degree of reextraction of their salts with 0.1 N NaOH is high, but their separation is not possible.

With Cyanex 923, acetic acid is coextracted with a high degree of extraction, lactic acid with a moderate degree and citric acid not at all. Therefore, their separation is possible. However, the reextraction of salts with NaOH is not possible, because stable emulsion is formed.

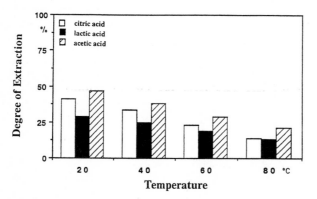

Fig. 4.4. Temperature dependence of extraction degrees of acetic, lactic and citric acids with Hostarex A 327-decanol system [238]

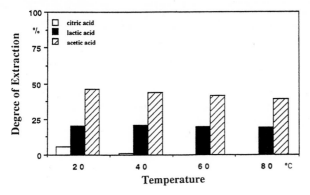

Fig. 4.5. Temperature dependence of extraction degrees of acetic, lactic and citric acids with Cyanex 932 system [238]

The reextraction of the free acids is rather difficult. With 2 M HCl, the reextraction from Hostarex A 327 and isodecanol solution gives the highest degree of reextraction of citric acid, a high degree of lactic acid, and a low degree of acetic acid. However, the total yield of the lactic acid is rather low (45%). With other inorganic acids the yields are still lower [238]. The reextraction of the free acids from Cyanex 923 solution with 2 M HCl the highest degree of reextraction is obtained for lactic acid, and low degrees for citric and acetic acid, but the total yield of the lactic acid is very low (max. 20%).

Yabannavar and Wang [242] recommended the reextraction of the free acid from Alamine 336/oleyl alcohol solution with conc. HCl. The obtained yield was fairly high (83%), however, the emulsion formed during the reextraction was very stable [243]. King et al. [241] used trimethyl amine (TMA) for the reextraction of carboxylic acids. The reextraction of lactic acid from Alamine 336/MIBK and Hostarex A 327 solutions respectively with TMA excess was possible with nearly 100% degree of reextraction. The decomposition of the acid/TMA complex was performed by vacuum distillation. The distillation residue was crystalline for succinic, malic, tartaric and citric acids, but not for lactic acid, because the latter formed a mixture of oligomers.

The reextraction of the salt and recovery of the free acid by electrodialysis using a bipolar membrane seems to be a simple practicable way. Still simpler is the direct recovery of the free acid from the filtered cultivation broth [228–232].

4.3 Extraction of Aromatic Carboxylic Acids

In contrast to aliphatic carboxylic acids, aromatic carboxylic acids are rarely formed by a microorganism, and so far no product is known that is commercially produced by fermentation. However, e.g. salicylic acid is formed by *Pseudomonas aeruginosa* from naphthalene by biotransformation [254].

In the following, the extraction of salicylic acid will be considered as a model for aromatic carboxylic acids. The extraction of salicylic acid from aqueous solutions by xylene and by a carrier (Amberlite LA-2, *N*-Lauryl-*N*-trialkyl-methylamine a secondary amine) in xylene was considered by Halwachs [244] and Halwachs and Schügerl [245] as well as by Schlichting et al. [246, 247]. The back-extraction was investigated by Haensel et al. [248, 249].

First of all, the extraction of salicylic acid from the aqueous phase by xylene and LA-2/xylene is dealt with [246]. Several processes have to be taken into account:

physical extraction of the nondissociated acid HS from the aqueous phase (index a) into the organic phase (index o)

$$HS_a \overset{P_c}{\rightleftharpoons} HS_o, \tag{4.8}$$

dissociation of the acid HS into acid anion S^- and hydroxonium ion in the aqueous phase

$$HS_a + H_2O_a \overset{K}{\rightleftharpoons} H_3O_a^+ + S_a^-, \tag{4.9}$$

extraction of the dissociated acid from the aqueous phase with the carrier amine A to form the complex AHS in the organic phase.

$$S_a^- + H_3O_a^+ + A_o \overset{K_e}{\rightleftharpoons} AHS_o + H_2O_a. \tag{4.10}$$

The distribution constant of the physical extraction is given by

$$K_p = \frac{[HS]_o}{[HS]_t} = P_c = \frac{1}{1 + 10^{pH - pK}}, \tag{4.11}$$

where P_c is the partition coefficient of the nondissociated acid

$$P_c = \frac{[HS]_o}{[HS]_a}. \tag{4.12}$$

[ ] Square bracket means equilibrium concentration and index t means total concentration.

pK = 2.97 at 20 °C.

The degree of extraction E (%) is given by

$$E = \frac{100}{1 + \dfrac{1 + 10^{pH - pK}}{P_c}}. \tag{4.13}$$

For the extraction with the amine carrier, the equilibrium constant is given by

$$K_e = \frac{[AHS]_o}{[H^+]_a [S^-]_a [A]_o}. \tag{4.14}$$

At first, only low acid concentrations are taken into consideration, where the dimer formation and the nonstoichiometric complex formation can be neglected. The component balances yield Eq. (4.15) for the equilibrium constant [246]:

$$K_e = \frac{[HS]_t - ([S^-]_a + [HS]_a)\left(1 + \dfrac{P_c}{1 + 10^{pH-pK}}\right)}{\left\{[A]_t - [HS]_t - ([S^-]_a + [HS]_a)\left(1 + \dfrac{P_c}{1 + (P_c/1 + 10^{pH-pK})}\right)\right\}[H^+]_a[S^-]_a}$$

(4.15)

and for the degree of extraction

$$E = 1 - \frac{[S^-]_a(1 + 10^{pH-pK})}{[HS]_t} \, 100$$

(4.16)

with

$$[S^-]_a = -0.5\left(\frac{[A]_t - [HS]_t}{R} + \frac{1}{K_e[H^+]_a}\right)$$

$$+ \left[0.25\left(\frac{[A]_t - [HS]_t}{K_e[H^+]_a} + \frac{1}{K_e[H^+]_a}\right)^2 + \frac{[HS]_t}{K_e[H^+]_aR}\right]^{0.5}$$ (4.17)

where

$$R = 1 + 10^{pK-pH} + P_c \, 10^{pK-pH}.$$

The partition coefficient P_c and the equilibrium constant K_e were determined to be:

$$P_c = 1.98$$

$$K_e = 2.69 \; 10^8 \; L^2 \, mol^{-2}.$$

By means of these values, the degree of extraction in the equilibrium was calculated as a function of the pH value in the aqueous phase and compared with the experiments (Fig. 4.6). There is a good agreement between calculated and measured data in the investigated concentration range

$$[A]_t = 0 \text{ to } 28.8 \text{ mM LA} - 2$$

where LA2 is a secondary amine (lauryl-trialkyl-methylamine, Amberlite LA-2).

The extraction kinetics were determined in a thermostatic stirred cell with on-line measurement of the solute concentration and the pH-values.

The reaction between acid and amine is instantaneous. Therefore, the rate-determining steps are the diffusion processes of the reactants from the bulk to the interface and the complex from the interface to the bulk. According to the two-film theory in the films of the aqueous and organic phases at the interface, concentration gradients exist. Only the gradient of H^+ can be neglected due to the high proton diffusivity (Fig. 4.7).

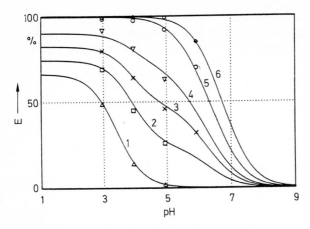

Fig. 4.6. Salicylic acid extraction: degree of extraction E as a function of the pH value at different amine concentrations in the equilibrium [246]

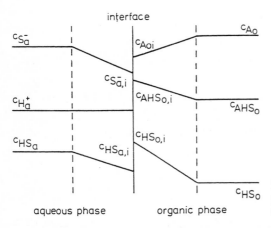

Fig. 4.7. Concentration profiles of components in the phases during salicylic acid extraction [246]

The transfer rate of the nondissociated acid due to the physical extraction is given by

$$-\frac{dC_{HSaP}}{dt} = a_s k_t (P_c C_{HSa} - C_{HSo}), \qquad (4.18)$$

where a_s is the specific interfacial area, C_{HSa} the concentration of the undissociated acid in the aqueous phase, C_{HSo} the concentration of the undissociated acid in the organic phase, index P means physical extraction and k_t is the overall mass transfer coefficient:

$$k_t = \frac{1}{\dfrac{1}{k_{HSo}} + \dfrac{P_c}{k_{HSa}}} \qquad (4.19)$$

k_{HSo} and k_{HSa} are the fractional mass transfer coefficients of undissociated acid in the organic and aqueous phases. P_c the partition coefficient of the undissociated acid at the interface i is assumed to be equal to the partition coefficient at equilibrium:

$$P_c = \frac{[HS]_{oi}}{[HS]_{ai}}$$

For the extraction with the amine carrier, the following fluxes can be defined:

for the complex

$$j_{AHSo} = k_{AHSo}(C_{AHSoi} - C_{AHSo}) \tag{4.20}$$

for the acid anion

$$j_{Sa^-} = (k_{Sa^-})(C_{Sa^-} - C_{Sa^-i}) \tag{4.21}$$

for the amine carrier

$$j_{Ao} = k_{Ao}(C_{Ao} - C_{Aoi}). \tag{4.22}$$

The three flux equations are connected by the mass action law:

$$K_e = \frac{C_{AHSoi}}{C_{Sa^-i} C_{Aoi} C_{Ha^+i}}. \tag{4.23}$$

In the steady state, the three fluxes are equal. By eliminating the interfacial concentrations, Eq. (4.24) is obtained.

$$-\frac{dC_{HSaR}}{dt} = k_{Sa}a_s \left\{ C_{Sa^-} + 0.5 \left(\frac{k_{Ao}}{k_{AHSo}K_e C_{Ha^+}} + \frac{k_{Ao}C_{Ao}}{k_{Sa^-}} - C_{Sa} \right) \right.$$
$$- \left[0.25 \left(\frac{k_{Ao}}{k_{AHSo}K_e C_{Ha^+}} + \frac{k_{Ao}C_{Ao}}{k_{Sa^-}} - C_{Sa^-} \right)^2 \right.$$
$$\left. \left. \times \frac{k_{Ao}C_{Sa^-}}{k_{AHSo}K_e C_{Ha^+}} + \frac{k_{Ao}C_{AHSo}}{k_{Sa^-}K_e C_{Ha^+}} \right]^{0.5} \right\} \tag{4.24}$$

In Eq. (4.24), index R means reactive extraction in contrast to index P (physical extraction).

Since the acid is removed from the aqueous phase by physical extraction (P) as well as by reactive extraction (R), the overall change of the acid concentration is given by the sum of Eqs. (4.18), (4.24).

$$-\frac{dC_{HSt}}{dt} = -\frac{dC_{HSaP}}{dt} - \frac{dC_{HSaR}}{dt}.$$

The overall mass transfer coefficients from the aqueous phase into the organic phase k_t and those of the back-extraction k_{tb} can be calculated from the measured concentration vs the time curves. They were measured for this system in a stirred cell, and the corresponding coefficients were calculated to be [246].

$$k_t = 0.47\ 10^{-3}\ \mathrm{cm\,s^{-1}}$$

$$k_{tb} = 3.00\ 10^{-3}\ \mathrm{cm\,s^{-1}}.$$

The fractional mass transfer coefficients in the organic phase were determined at pH 12 in the aqueous phase at which the mass transfer resistance in the aqueous phase is completely eliminated by shifting the reaction front to the interface (See Sect. 2.3.4). In the same system, the fractional mass transfer coefficients for the complex k_{AHSo} and the amine carries k_{Ao} were indentified to be

$$k_{AHSo} = 0.9\ 10^{-3}\ \mathrm{cm\,s^{-1}}$$

$$k_{Ao} = 1.25\ 10^{-3}\ \mathrm{cm\,s^{-1}}.$$

The knowledge of k_t, k_{AHSo} and k_{Ao} allows one to calculate the fractional mass transfer coefficient k_{Sa^-} for the acid anion by means of Eq. (4.24). k_{Sa^-} was identified for the above system as

$$k_{Sa^-} = 1.5\ 10^{-3}\ \mathrm{cm\,s^{-1}}.$$

In Fig. 4.8, the concentration vs the time curves were calculated with these coefficients for the extractions with different amine concentrations and compared with the measured data. The agreement between the calculated and measured curves for physical extraction (by xylene) as well as for reactive extraction (by amine carrier in xylene) is satisfactory. Similar agreements were also found for the back-extractions.

This example indicates that equilibrium and kinetic data can be described by Eqs. (4.15), (4.24) for physical and reactive extractions of carboxyl acids.

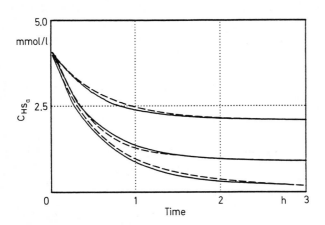

Fig. 4.8. Kinetics of salicylic acid extraction at different amine (LA-2) concentrations: salicylic acid concentration in the aqueous phase as a function of the extraction time [246]
curve 1: LA2 = 0 m mol/l
curve 2: LA2 = 2.4 m mol/l
curve 3: LA2 = 6.8 m mol/l
—— measured
– – – –calculated

However, here this system was examined for low acid concentrations, in order to neglect the dimer formation and the super-stoichiometric extraction. At higher acid concentrations, these phenomena cannot be neglected any longer. Haensel et al. [248] investigated the physical and reactive extraction of salicylic acid at high concentrations. They took into account dimerisation as well as superstoichiometric extraction. Here their model should serve as an example.

Equations (4.25)–(4.28) describe the equilibrium for acid dissociation, monomer equilibrium between the phases, dimerisation and complex formation:

$$HS_a + H_2O \overset{K}{\rightleftharpoons} H_3O^+ + S_a^- \tag{4.25}$$

$$HS_a \overset{P_c}{\rightleftharpoons} HS_o \tag{4.26}$$

$$HS_o \overset{K_{di}}{\rightleftharpoons} (HS)_{2,o} \tag{4.27}$$

$$HS_o + A_o \overset{K_1}{\rightleftharpoons} AHS_o \tag{4.28a}$$

$$HS_o + AHS_o \overset{K_2}{\rightleftharpoons} A(HS)_{2,o} \tag{4.28b}$$

$$HS_o + A(HS)_{m-1,o} \overset{K}{\rightleftharpoons} A(HS)_{m,o} \tag{4.28c}$$

From the mass balances of acid and amine the superstoichiometric factor β can be calculated:

$$\beta = \frac{\sum_{j=1}^{n} \{j(\prod_{i=1}^{j} k_i)(P_c[HS_a])^j\}}{1 + \sum_{j=1}^{n} \{(\prod_{i=1}^{j} K_i)(P_c[HS_a])^j\}} \tag{4.29a}$$

or

$$\beta = \frac{[HS_{o,t}] - P_c[HS_a] - 2P_c^2 K_{di}^*[HS_a]^2}{[A_t]}. \tag{4.29b}$$

β describes the ratio of acid present as a complex in the solvent phase to the total amount of carrier.

$\quad \beta > 1$ sub stoichiometric carrier complex $\qquad\qquad \beta > 1$

$\quad \beta = 1$ stoichiometric carrier complex $\qquad\qquad\quad \beta = 1$

$\quad \beta < 1$ superstoichiometric carrier complex $\qquad\quad \beta < 1$

The reduced partition coefficient P_r is defined as follows:

$$P_r = \frac{\sum_{j=1}^{n} \{j(\prod_{i=1}^{j} k_i)(P_c^j[HS_a]\}^{j-1}\}}{1 + \sum_{j=1}^{n} \{(\prod_{i=1}^{j} K_i)(P_c[HS_a])^j\}} [A_t] + P_c + 2P_c^2 K_{di}^*[HS_a] \tag{4.30}$$

Limiting cases of Eq. (4.30):
for $[A_t] = 0$ (no carrier) it is derived from Eq. (4.30)

$$P_r = P_c + 2P_c^2 K_{di}^*[HS_a], \tag{4.31}$$

this holds true for physical extraction.

For low acid concentration, $[HS_a] \rightarrow 0$, it is derived from Eq. (4.30)

$$P_r \approx K_1 P_c [A_t] + P_c. \tag{4.32}$$

Equation (4.32) allows the determination of the order of magnitude of the stability constant of the stoichiometric acid-carrier complex, if P_r is a linear function of the carrier concentration. For the system investigated by Haensel et al. [248] this linearity holds true. For this system the equilibrium constants were identified to be: $K = 10^{-2.97}$, $P_c = 1.9$, $K_{di}^* = 21.6 \, L \, mol^{-1}$, $K_1 = 2.4 \, 10^5 \, l \, mol^{-1}$, $K_2 = 1.8 \, 10^5 \, l \, mol^{-1}$.

The kinetics of the extraction can be determined in a small stirred cell. The solute concentration is measured as a function of the extraction time. To estimate the kinetics of the extraction the two-film theory can be used again. In the following, the kinetic parameters of the back-extraction of salicylic acid from the xylene solvent in absence and in presence of amine carrier into the aqueous phase will be taken into consideration according to Haensel et al. [248].

The concentration profiles of the components at the interface are shown in Fig. 4.9. The total flux in the organic donor phase is given by

$$j_{HSo,t} = j_{HSo} + j_{(HS)2,o} = k_{o,t} (C_{HSo,t} - C_{HSo,t,i}), \tag{4.33}$$

where HS_o and $(HS)_{2,o}$ are the monomer and dimer acids in the organic phase which are simultaneously transferred to the interface.

In the aqueous receiver phase, the nondissociated and dissociated acids are transferred simultaneously. The two fluxes are coupled with the equilibrium condition:

$$HS_a + OH_a^- \overset{K_B}{\rightleftharpoons} S_a^- + H_2O. \tag{4.34}$$

The equilibrium constant was determined to be $K_B = 10^{11} \, l \, mol^{-1}$ at $pH = pK$.

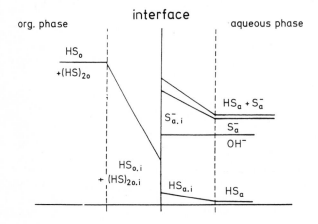

Fig. 4.9. Concentration profiles of components at the interface during the salicylic acid extraction [249]

This reaction is an instantaneous bimolecular reaction. Its reaction front shifts towards the interface with increasing OH^- concentration. At a high pH, the reaction front is at the interface, and the mass transfer resistance in the aqueous film is completely eliminated.

The total flux in the aqueous receiver phase is given by:

$$j_{HSa,t} = j_{HSa} + j_{Sa^-} = k_a(1 + K_B C_{OH^-})(C_{HSa,i} - C_{HSa}) \qquad (4.35)$$

The total flux across the interface is given by Eq. (4.36) when assuming an equilibrium at the interface:

$$-j_{HSo,t} = k_t(C_{HSo,t} - D_c C_{HSa}), \qquad (4.36)$$

where k_t is the overall mass transfer coefficient and D_c is the distribution ratio of the acid

$$D_c = \frac{[HS_{o,t}]}{[HS_{a,t}]}$$

$$k_t = \left[\frac{1 + B}{k_o + k_{di}B} + \frac{D_c}{k_a(1 + K_B C_{OH^-})} \right]^{-1} \qquad (4.37)$$

$$B = 2 K_{di}^* C_{HSo}. \qquad (4.38)$$

k_o and k_a are the mass transfer coefficients of the monomer solute in the organic and aqueous phases, and k_{di} the mass transfer coefficient of the dimer in the organic phase.

At a low acid concentration, the dimer fraction can be neglected (e.g. 97% of the acid prevails as monomer at 1 mM concentration). Thus $B \to 0$, $D_c \to P_c$ and Eq. (4.37) can be reduced to Eq. (4.39)

$$k_t = \left[\frac{1}{k_o} + \frac{P_c}{k_a(1 + K_B C_{OH^-})} \right]^{-1} \qquad (4.39)$$

It is possible to determine k_o and k_a separately. At pH 1, $K_B C_{OH^-} \to 0$ as well as Eq, (4.40)

$$k_t = \left[\frac{1}{k_o} + \frac{P_c}{k_a} \right]^{-1} \qquad (4.40)$$

and at pH 12 Eq. (4.41)

$$k_t = k_o \qquad (4.41)$$

hold true. From eqs. (4.40), (4.41) k_o and k_a can be calculated.

In practice, the acid concentration can be high enough to have a considerable fraction of dimers in the organic phase. The mass transfer coefficient of the dimer can be determined at pH 12, where the following relationship is valid for the overall solute flux in the organic phase:

$$-j_{HSo,t} = \left(\frac{1}{1 + B} k_o + \frac{B}{1 + B} k_{di} \right) C_{HSo,t} \qquad (4.42)$$

As already mentioned, at this high pH-value the mass transfer resistance of the acid in the aqueous phases can be neglected, thus, if the dimer fraction is known and k_o was determined at a low solute concentration of pH 12, k_{di} can be calculated by means of Eq. (4.42).

In the presence of a carrier in the organic phase the transfer of the acid-carrier complex has to be considered too. The overall flux of acid through the interface at a high pH is given by

$$j_{HSo,t} = k_t C_{HSo,t},$$ (4.43)

where k_t is defined by Eq. (4.44)

$$k_t = k_{o,t} = \frac{1}{1+B+C} k_o + \frac{B}{1+B+C} k_{di} + \frac{C}{1+B+C} k_k$$ (4.44)

where k_k is the mass transfer coefficient of the complex in the organic phase.
The variable C is a function of the carrier concentration:

$$C = \frac{K_1 + 2K_1 K_2 [HS_o]}{1 + K_1 [HS_o] + K_1 K_2 [HS_o]^2} [A_t].$$ (4.45)

The dimer transfer coefficient k_{di} can be calculated from

$$k_{di} = k_t \frac{1+B}{B} - k_o \frac{1}{B},$$ (4.46)

At a low solute concentration, the dimer fraction can be neglected, therefore $B \to 0$ and Eq. (4.45) can be reduced to

$$k_t = k_{o,t} = \frac{1}{1+c} k_o + \frac{1}{1+c} k_k.$$ (4.47)

At pH 1, the acid exists in an undissociated state, and it is extracted only by the solvent (physical extraction). With increasing acid concentration, k_t becomes smaller due to the higher dimer fraction in the organic phase. With increasing extraction time, k_t increases and approaches the pure monomer transfer coefficient. Under these conditions, (low solute concentration) Eq. (4.46) also allows one to calculate k_k, the mass transfer coefficient of the acid-carrier complex.

The film model of Haensel et al. presented here makes it possible to determine all of the fractional mass transfer coefficients of such a complex system experimentally, if one also performs mass transfer measurements at pH 1 and pH 12.

It is well known that the consecutive reaction is able to accelerate the mass transfer. The acceleration can be characterized by the ratio of the solute fluxes with and without chemical reaction:

$$\Phi = \frac{- j_{HSo} \text{ (with consecutive reaction)}}{- j_{HSo} \text{ (without consecutive reaction)}},$$ (4.48a)

where Φ is an enhancement factor.

According to this definition for a reversible bimolecular instantaneous reaction, the following relationship holds true [244]:

$$\Phi = \frac{K_B C_{OH^-}}{1 + \dfrac{k_a}{P_c k_o}(1 + K_B C_{OH^-})}. \tag{4.48b}$$

For a high pH, $K_B C_{OH^-} \to \infty$, the maximum enhancement factor can be attained:

$$\Phi_{max} = 1 + \frac{P_c k_o}{k_a}. \tag{4.49}$$

With an increasing pH from pH 1 on, the enhancement factor increases due to the spontaneous dissociation of the acid at the interface during the transfer into the aqueous phase. Because of this, the mass transfer at pH 3 is already twice as high as that at pH 1. At about pH 12 the mass transfer attains its maximum value. The maximum enhancement factor in the case of salicylic acid is 5.

However, the fluid dynamic conditions in stirred cells with a flat interface are fairly different from those in commercial equipment, in which one of the phases is dispersed into the other one, and the mass transfer occurs between the continuous and the dispersed phases. To work out the fundamentals for this type of mass transfer, investigations can be carried out with single droplets according to techniques discussed in Chapter 2.4.2.

Since the most advanced investigation method for mass transfer between single droplets and their environment is the modified liquid scintillation technique, in the following some results are presented on such investigations carried out with salicylic acid as solute by Haensel et al. [249]. Although such research work has already been done with acetic acid [251, 252], this aromatic carboxylic acid was chosen to establish a connection between the investigations in stirred cells discussed above and single drop measurements.

The measuring technique has already been discussed in Sect. 2.4.2. The measured concentration vs the time curves are evaluated by means of a simple quasi-steady state film model which yields the instantaneous flux j_{HSd} (index d means droplet) and the overall mass transfer coefficient k_t.

$$j_{HSd} = -\frac{1}{F} V_d \frac{dC_{HSd}}{dt} \tag{4.50}$$

$$k_t = -\frac{1}{F} V_d \frac{dC_{HSd}}{dt} \frac{1}{C_{HSd}}, \tag{4.51}$$

where F is the surface area of the droplet V_d the volume of the droplet, and C_{HSd} the concentration of the acid in the droplet.

The use of these simple relationships are possible, since the solute concentration int he continuous aqueous bulk phase equals zero.

In order to investigate the fluid dynamic effects on the mass transfer from the droplet into the continuous aqueous phase, the droplet volumes were varied from 10 to 100 μl and the droplets were fixed on the tip of a hypodermic needle or were freely suspended in a steady state or pulsed flow of the continuous aqueous phase. With increasing droplet size, the droplet Reynolds number Re_d varies regarding the continuous phase, including the droplet diameter d_d and the relative velocity of droplet u_d. The droplet Reynolds number

$$Re_d = \frac{d_d u_d}{v_c}$$

was varied between 100 and 300, where v_c is the kinematic viscosity of the continuous phase. Only below 50 μl does u_d depend on the droplet volume. To investigate the influence of the fluid dynamics on the physical transfer of the undissociated acid, the pH-value was reduced by at least two units below its pK = 2.97. By using hydrochloric acid, a pH of 0.8 was maintained. Under these conditions, the dissociation of the salicylic acid is less than 1% of its overall concentration. Droplet volumes of 10, 25, 50 and 100 μl were used.

The droplet formation, its separation from the hypodermic needle and its rise into the measuring position took about 1 s. The solute concentration reduces within the first 50 to 100 s to about 80% of its original value; at the same time, the mass transfer rate also diminishes. Thereafter, the mass transfer rate approaches a constant value depending on the droplet volume.

The increase of the initial acid concentration from 11.1 to 45.1 mM reduces the mass transfer rate and the mass transfer coefficient due to the increase of the dimer fraction in the organic droplet phase. The mass transfer coefficient increases considerably from the fixed droplet ($Re_d = 0$) and freely suspended droplet in a steady state flow ($Re_d = 190$) and in a pulsed flow ($Re_d = 190$–220).

The increase of the pH from acidic (pH 1) to the neutral (pH 7) range causes an increase in the mass transfer coefficient by a factor of 2.5 due to the spontaneous dissociation of the acid. In the pH range from 10 to 12, the coefficient steeply increases by the shift of the reaction front in the direction of the interface. To eliminate the mass transfer resistance in the aqueous film and determine the fractional mass transfer coefficient k_d in the droplet phase, 0.01 M NaOH was used to maintain pH 12 during the extraction process. In good agreement with the stirred cell measurements, the k_t values for reactive extraction are higher by a factor of four to five than those without chemical reaction. With increasing solute concentration, k_t falls due to the increase of the dimer fraction.

The knowledge of k_d permits the calculation of the fractional mass transfer coefficient k_c for the continuous phase. k_c was calculated by Eq. (4.52).

$$k_t = \left[\frac{1}{k_d} + \frac{P_c}{k_c} \right] - 1 \tag{4.52}$$

k_d was determined at pH 12, where $k_t = k_d$.

In Table 4.21a, b the mass transfer coefficients for different chemical conditions are compared. In order to avoid the influence of dimer transfer in Table 4.21, only k_t values were used which were determined at high extraction times at which dimer fraction was below 5%.

In Table 4.22a, b the influence of the hydrodynamic conditions are shown.

Tables 4.21, 4.22 show the significant influences of the pH, acid concentration, droplet volume and fluid-dynamic conditions on the mass transfer coefficients.

The influence of the pH on the enhancement factor is shown in Fig. 4.10. After its gradual increase between pH 1 and pH 10, the factor exhibits a strong increase between pH 10 and 12, where the complete elimination of the mass transfer resistance due to the reaction between acid and NaOH in the continuous phase occurs. No significant change of the enhancement factor can be observed above pH 12 (not shown in Fig. 4.10). The influence of the droplet volume on the mass transfer coefficient is shown in Fig. 4.11. The reduction of the coefficients with an increasing V_d is caused by the change of the droplet shape [249]. With increasing droplet volume the droplet shape deviates more and more from the spherical shape. Large flat droplets have a higher flow resistance due to the vortex trail formation on the downstream side of the droplet, which also reduces the mass transfer on that side. For large droplets, only the surface area on the upstream side of the droplet is active in the mass

Table 4.21. Overall and fractional mass transfer coefficients on freely suspended droplets for different chemical conditions [249]

a) Influence of pH on the continuous phase:
acid concentration 45 mM, $V_d = 20\ \mu l$, $a_d = 17.8\ cm^{-1}$, $Re_d = 190$, $D_c = 1.9$

pH	$k_t 10^3$ (cm/s) physical	$k_t 10^3$ (cm/s) reactive extraction	$k_d 10^3$ cm/s reactive	$k_c 10^3$ (cm/s) physical
1	0.6	—	—	—
3	—	1.2	—	—
7	—	1.5	—	—
10	—	1.6	—	—
12	—	—	2.7	1.5
13	—	—	2.8	1.5

b) Parameter initial acid concentration C_{HSdo}
$V_d = 25\ \mu l$, $a_d = 16.52\ cm^{-1}$, $Re_d = 190$, $D_c = 1.9$

C_{HSdo}	$k_t 10^3$ (cm/s) physical	$k_d 10^3$ (cm/s) reactive extraction	$k_c 10^3$ cm/s physical
11.1	0.65	2.85	1.6
22.3	0.6	2.80	1.5
45.1	0.58	2.70	1.4

Table 4.22. Overall and fractional mass transfer coefficients on single droplets under different hydrodynamic conditions [249]

a) Influence of the droplet volume
$C_{HSdo} = 45\,mM$, $D_c = 1.9$

V_d (µl)	a_d (cm^{-1})	Re_d (1)	$k_t 10^3$ (cm/s) physical	$k_d 10^3$ cm/s reactive extraction	$k_c 10^3$ (cm/s) physical
10	22.4	115	0.7	3.2	1.7
20	17.8	165	0.65	2.85	1.6
25	16.5	190	0.6	2.8	1.4
50	13.1	252	0.53	3.25	1.2
100	10.4	317	0.49	2.17	1.2

b) Influence of the hydrodynamic conditions
$C_{HSdo} = 45\,mM$, $V_d = 25$ microlitres, $a_d = 16.5\,cm^{-1}$, $D_c = 1.9$

Droplet	Re_d (1)	$k_t 10^3$ (cm/s) physical	$k_d 10^3$ (cm/s) reactive extraction	$k_c 10^3$ (cm/s) physical
fixed	0	0.38	0.85	1.3
freely suspended	190	0.58	2.70	1.4
pulsated	190–220	0.82	3.80	1.98

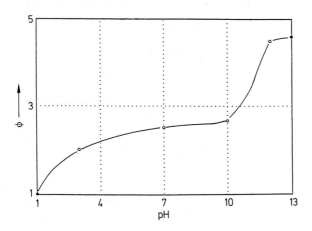

Fig. 4.10. Enhancement factor Φ as a function of the pH value for salicylic acid extraction [249]

transfer. Therefore, with increasing droplet size, the droplet surface area which participates actively in the mass transfer, reduces more and more to the upper half of the surface area.

The fractional mass transfer coefficients evaluated in this way can be compared with the predictions of the theoretical models.

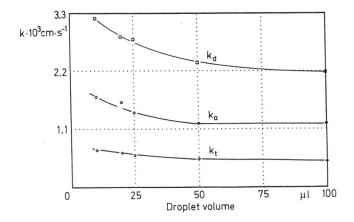

Fig. 4.11. Mass transfer parameter as a function of the droplet volume. Overall and fractional mass transfer coefficients [249]

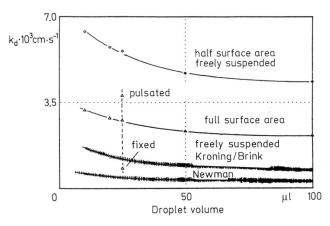

Fig. 4.12. Comparison of measured fractional mass transfer coefficients k_d in the droplet with model calculations [249]

In Fig. 4.12, the measured fractional mass transfer coefficients k_d in the droplet are compared with model predictions at low solute concentrations at which the dimer transfer is negligible.

The transfer coefficients in droplets fixed to the tip of a hypodermic needle are between the predictions of the Newman model ([23] in Chapter 2), which assumes a pure diffusive transfer, and the Kronig-Brink model ([24] in Chapter 2), which takes into consideration laminar circulation within the droplet. The coefficients in freely suspended droplets are above the predictions of the Kronig-Brink model. The same is valid for the coefficients measured in the pulsated flow. However, they are much lower than the Handlos-Baron model ([25] in Chapter

2) predicts, which assumes a turbulent circulation movement within the droplets. The predictions of this model are not shown in Fig. 4.12, since they fall outside of the range of this figure. For the calculation of k_d, the full surface area as well as the half surface area of the droplets were used, since it is not possible to calculate the fraction of the surface area which participates actively in the mass transfer. The actual fractions are between the full surface area (for small droplets) and half the surface area (for large droplets).

In Fig. 4.13, the measured fractional coefficients (k_c) in the continuous phase are compared with the predictions of the Higbie model ([18] in Chapter 2) which assumes surface renewal with a deterministic residence time of the fluid elements at the interface. Since the mass transfer occurs from the droplet to the continuous phase, it was assumed that only the upstream side of the spherical droplet participates in the mass transfer [245].

A comparison of the measured and predicted coefficients show large differences, k_c values measured with fixed and freely suspended droplets are considerably lower than those predicted by the Higbie model. Only the coefficients evaluated in the pulsated flow come close to these theoretical values.

In presence of the amine carrier, the acid prevails in the droplet at least in three different species: monomer, dimer and acid-amine complex. In order to investigate the mass transfer in the droplet, the measurements were performed at pH 13, at which the mass transfer resistance of the aqueous phase disappears. Since the mass transfer coefficients of the monomer and dimer are higher than that of the complex, they are extracted faster from the droplet, thus the complex fraction gradually increases. This causes a reduction of k_t with the time. After 150 s, acid monomer and dimer have practically disappeared from the droplet, thus the mass transfer of the complex controls the extraction process alone. This yields a constant k_t value for the rest of the extraction (Fig. 4.14), which

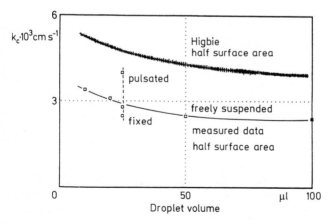

Fig. 4.13. Comparison of measured fractional mass transfer coefficients k_c in the continuous phase with model calculations according to Higbie [249]

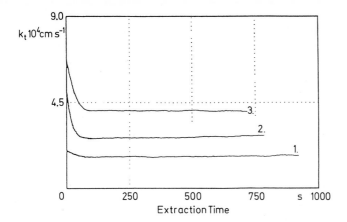

Fig. 4.14. Overall mass transfer coefficient k_t as a function of time at high salicylic acid concentration [248]. Droplet volume 25 μl, pH < 1.0, T = 22 °C, 1, fixed droplet, Re = 0; 2, freely suspended droplet, Re = 190, 3, pulsated droplet, Re = 190–220

corresponds to the ratio of the actual acid-to-amine concentration 2:1. This proves that a superstoichiometric complex was formed.

The increase of the amine carrier concentration reduces k_d because of the low mass transfer coefficient of the complex from 2.8×10^{-3} cm/s (no amine carrier) to 1.1×10^{-3} cm/s (5 mM carrier), 1.05×10^{-3} cm/s (11 and 22 mM carrier) and 0.95×10^{-3} cm/s (45 mM carrier).

Mass transfer on freely suspended single droplets are closer to that on droplet swarms in commercial equipment than mass transfer in stirred cells. However, with single droplets, the coalescence/redispersion effects cannot be covered. Therefore, investigations on laboratory-scale and pilot-plant-scale extractors are also necessary.

In order to include these effects, the extraction of salicylic acid from a buffer-free continuous aqueous phase by a variable amount of amine carrier in xylene solvent in a laboratory-scale pulsed sieve plate column (Fig. 4.15) and the back extraction by an 0.01 N NaOH solution in the same equipment are considered [247]. The column data are given in Table 4.23 and the operating parameters in Table 4.24.

The extraction was performed by an LA-2 amine carrier in xylene as the dispersed phase from the buffer-free continuous aqueous phase. The back-extraction was carried out by 0.01 N NaOH continuous phase from the LA-2 amine carrier/xylene droplet phase.

The holdup of the dispersed phase ε was measured by two different methods: by the γ-transmission technique using a 100 Ci ^{137}Cs source and by a 7.62×7.62 cm NaJ (Tl) detector and by draining the two-phase system and measuring the volumes of the two phases.

The droplet-size distribution was evaluated by flash photography and a semiautomatic image analyser. At least 800 droplets were used for the evaluation of the droplet diameter distributions. The droplet Sauter diameter d_{32} was

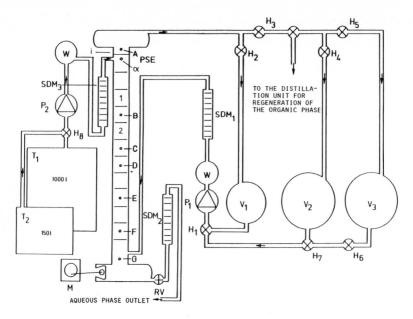

Fig. 4.15. Pulsed sieve plate column for extraction of salicylic acid with and without amine extractants in xylene [247]

Table 4.23. Data of laboratory-scale pulsed sieve plate extraction column [247]

Inner column diameter	100 mm
Overall height	3330 mm
Overall volume	21.9 l
Active length	2235 mm
Active volume	18.51 l
Number of trays	20
Tray distance	100 m
Hole diameter on perforated plate	2 mm
Free surface area	20%
Stroke frequency	0–160 strokes min^{-1}
Stroke height	0–22 mm

Table 4.24. Operating parameters of the extraction of salicylic acid in a laboratory scale pulsed sieve-plate column [247]

Continuous phase	Solute conc (mM)	V_o (l/h)	V_a (l/h)	f (min^{-1})	a_d (cm^{-1})	C_A (%)
aqueous (extraction)	10	70–120	100	100	0.9	2.4–28.8
aqueous (back-extraction)	8	70–120	100	80–120	0·9	2.4–28.8

calculated from the droplet diameter distribution, and the specific interfacial area was evaluated from

$$a_d = \frac{6\varepsilon}{d_{32}}.$$ (4.53)

The distribution of the residence times of both phases was measured by a dye-tracer pulse technique and evaluated by the moment method [247]. The mean residence times and the number of equivalent stages of the cascade model were determined to simulate the extraction process in the laboratory-scale extractor. The longitudinal solute concentration profiles were also measured in the continuous aqueous phase in the extraction column. The extraction kinetics evaluated in a stirred cell [246] and the cascade model with the model parameters evaluated by separate measurements were used for the mathematical simulation of the extraction process in the column [247].

In Figs. 4.16, 4.17 the measured longitudinal solute concentration profiles are shown with xylene (Fig. 4.16) and carrier-xylene (Fig. 4.17) solvents at different volumetric flow rates (V_o) of the organic phase. With increasing V_o and in presence of the LA-2 carrier, the extraction process is accelerated.

Since a buffer-free aqueous phase was used, the pH value increased with the degree of extraction in the aqueous phase. Therefore, the stroke frequency had only a slight influence on the degree of extraction with xylene as well as with carrier/xylene solvents, because at the end of the column, the pH of the aqueous phase shifted to higher values due to the increased extraction degree. This caused a shift of the extraction equilibrium to the back-extraction.

With increasing carrier concentration, the degree of extraction reached a maximum at an equimolecular ratio of acid to amine (Fig. 4.18), because at high amine concentrations, the pH increased and, as a consequence of this, the

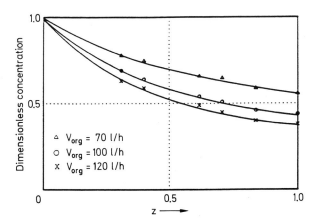

Fig. 4.16. Extraction of salicylic acid with xylene: longitudinal profiles of the dimensionless concentration at different volumetric flow rates of the organic phase [247]. fa = 1.5 cm s^{-1}

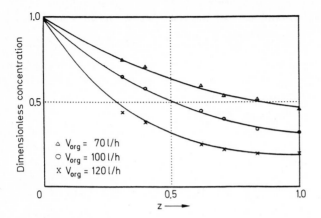

Fig. 4.17. Extraction of salicylic acid with amine extractant (LA-2) in xylene: longitudinal profiles of the dimensionless concentrations at different volumetric flow rates of the organic phase [247]. fa = 1.5 cm s^{-1}

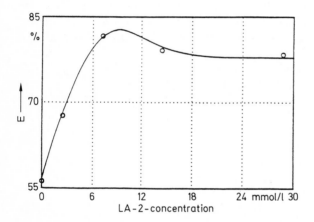

Fig. 4.18. Extraction of salicylic acid: degree of extraction as a function of the amine (LA-2) concentration [247]. $V_o = V_a = 100 \, \mathrm{l \, h^{-1}}$, $C_{HS} = 10 \, \mathrm{mM}$, fa = 1.5 cm s^{-1}

equilibrium shifted again toward the back-extraction [247]. Also the holdup of the organic phase exhibits a maximum at the equimolecular ratio of acid to amine. The Sauter droplet diameter d_{32} becomes larger with increasing amine concentration due to the reduction of the interfacial tension. With rising stroke frequency (f), the Sauter diameter is reduced because of the increased specific power input.

The computer simulation of the extraction with xylene agreed well with the measurements (Fig. 4.19). However, the calculated and measured longitudinal

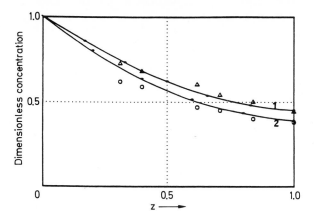

Fig. 4.19. Calculated (*curves*) and measured (*symbols*) longitudinal profiles of dimensionless concentrations of salicylic acid during extraction by xylene at different stroke intensities [247]. $\triangle$ fa = 1.2 cm s^{-1}, $\bigcirc$ fa = 1.8 cm s^{-1}

concentration profiles disagree if the extraction is performed by a carrier/xylene solvent. This can be attributed to the model parameters of the extraction column, which were measured with a xylene/water system. Obviously the presence of the amine in xylene strongly influences the fluid-dynamic parameters of the two-phase system.

4.3.1 Symbols for Sect. 4.3

A_o	amine in the organic phase
$[A]_t$	$[A]_o + [AHS]_o$
a_d	specific interfacial area in the column
a_s	specific interfacial area
AHS_o	acid-amine complex in the organic phase
C	variable (Eq. (4.45))
C_{Ao}	amine concentration in the organic phase
C_{AHSo}	acid-amine complex concentration in the organic phase
C_{HSa}	concentration of HS in the aqueous phase
C_{HSd}	concentration of HS in the droplet
C_{HSo}	concentration of HS in the organic phase
C_{OH^-}	concentration of OH$^-$
C_{Sa^-}	concentration of S$_a^-$ in the aqueous phase
D_C	distribution ratio of acid (Eq. (4.36))
E	degree of extraction
F	surface of the droplet
f	frequency of pulsated perforated plate column
H_a^+	proton in the aqueous phase

HS_a	acid in the aqueous phase
HS_o	acid in the organic phase
$[HS]_{at}$	$[HS]_a + [S^-]_a$
$[HS]_{ot}$	$[HS]_o + 2[(HS)_2]_o$
$[HS]_t$	$[S^-]_a + [HS]_a + [HS]_o + 2[(HS)_2]_o + [AHS]_o$
$(HS)_{2o}$	acid dimer in the organic phase
j_{Ao}	flux of A_o
j_{AHSo}	flux of AHS_o
J_{Ha^+}	flux of H_a^+
j_{HSa}	flux of HS_a
j_{HSd}	instantaneous flux (in droplet) (Eq. (4.50))
j_{HSo}	flux of HS_o
j_{Sa^-}	flux of S_a^-
K	dissociation coefficient
K_B	equilibrium constant (Eq. (4.34))
K_{di}	equilibrium constant of the dimer formation in the organic phase
K_e	equilibrium constant
K_p	distribution constant of physical extraction
K_1	equilibrium constant of the acid-amine complex formation
K_2	equilibrium constant of the formation of $A(HS)_{2'o}$ complex
k_{Ao}	mass transfer coefficient of A_o
k_{AHSo}	mass transfer coefficient of AHS_o
k_a	mass transfer coefficient of the monomer solute in the aqueous phase
k_{di}	mass transfer coefficient of $(HS)_2$
k_{HSa}	mass transfer coefficient of HS_a
k_{HSo}	mass transfer coefficient of HS_o
k_k	mass transfer coefficient of the complex in the organic phase
k_o	mass transfer coefficient of the monomer solute in the organic phase
k_{Sa^-}	mass transfer coefficient of S_a^-
k_t	overall mass transfer coefficient (in droplet) (Eq. (4.51))
k_t	Eq. (4.19) overall mass transfer coefficient
k_{tb}	overall mass transfer coefficient of back extraction
OH^-	OH^- anion
P_c	partition coefficient
P_r	reduced partition coefficient (Eq. (4.30))
Re_d	droplet Reynolds number
S_a^-	acid anion in the aqueous phase
V_a	throughput of the aqueous phase
V_d	droplet volume
V_o	throughput of the organic phase
β	stoichometric factor (Eq. (4.29))
Φ	enhancement factor (Eq. (4.48))

4.4 Extraction of Amino Acids

Several amino acids are produced commercially by enzymatic biotransformation, microbial biosynthesis or protein hydrolysis [255]. Amino acids produced by biosynthesis (e.g. lysine) or enzymatic biotransformation (e.g. methionine) are present in high concentration and the number and concentration of accompanying components are low. Their recovery is easy. On account of their identical charged groups, the recovery of particular amino acids from protein hydrolysates is rather difficult.

Separation of amino acids of basic character (lysine, arginine, histidine) from those of acidic character (aspartic acid, glutamic acid) by extraction is possible. Likewise, the separation of amino acids with polar side groups (glycine, serine, threonine, cysteine, tyrosine, glutamine) from those with a nonpolar side group (alanine, valine, leucine, isoleucine, proline, phenylalanine, tryptophan, methionine). However, the separation of amino acids with similar chemical character and solubility is only possible at the present on industrial scale by chromatography.

Amino acids have two or three charged groups, form zwitter ions in the neutral range, and except for threonine all of the bifunctional acids have pK_1 (COO^-) values between 2 and 3, and pK_2 (NH_3^+) values between 9 and 10.

This also holds true for trifunctional amino acids, with the exception of cystein with $pK_1 = 1.71$ and $pK_2 = 10.787$, and histidine with $pK_1 = 1.82$.

The trifunctional amino acids differ only in their pK_3 (side group) values: 3.36 (aspartic acid), 4.25 (glutamic acid), 6.0 (histidine), 8.33 (cystein), 10.07 (tyrosine), 10.53 (lysine) and 12.40 (arginine).

Due to the amino ($-NH_2$) and the carboxyl ($-COOH$) groups of the amino acids, they behave as cations at a low pH, as anions at a high pH, and they are of a zwitterionic character at an intermediate pH. Though they have no net charge in the intermediate pH range, because cationic and anionic groups neutralize each other, their solubility in nonpolar solvents is very low. Their solvent extraction with nonpolar extractants is not possible. Their solvation with carbon-bonded, oxygen-bearing extractants is inefficient.

Extraction with phosphorus-bonded, oxygen-bearing extractants by proton transfer or with high-molecular-weight quaternary aliphatic amines is effective. E.g., at a low pH (2 to 3), the amino acid has a positive net charge. Consequently, a cation carrier (e.g., di-(2-ethylhexyl)phosphoric acid (D2EHPA) can be used as extractant [256].

At a high pH (10–12), the amino acid has a negative net charge. An anion carrier (e.g., tri-capryl ammonium chloride (Aliquat 336, Henkel Co) is a suitable extractant [257].

Behr and Lehn [258] were the first ones to report on the transport of amino acids and dipeptides through a bulk toluene membrane separating two aqueous phases by means of protonation, deprotonation processes and countertransport of an inorganic ion. An amino acid in the carboxylate form RCH(NH_2)-COO-may be transferred from an alkaline solution (0.1 N KOH) to a membrane, when a positively charged, highly lipophilic carrier (quaternary ammonium chloride, e.g., Aliquat 336) is added to the membrane. As the amino acid ammonium salt reaches the acid phase (0.1 N HCl) on the other side of the membrane, it is extracted into this phase by protonation: $R\text{-}CH\text{-}(COOH)NH_3^+$. The exchanged chloride anion is transported countercurrently through the membrane phase into the alkaline phase (Fig. 4.20a).

When starting with the cation $(R\text{-}CH\text{-}(COOH)NH_3^+)$ in the acid phase and using a negatively charged carrier (e.g., dinonylnapthalenesulfonate, $DNNS^-$), the amino acid sulfonate ion pair is transported through the membrane and extracted into the alkaline solution (0.1 N KOH) as carboxylate. The exchanged potassium ion is transported countercurrently into the acid phase (Fig. 20b).

Behr and Lehr investigated the transported of several amino acids with the same three-phase system and found a transport of amino acids against Cl^- or

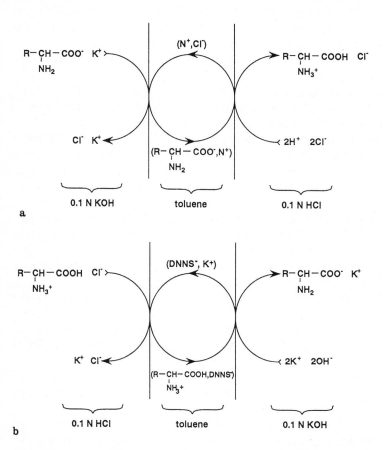

Fig. 4.20a, b. Transport of amino acids through a toluene membrane layer: **a** from basic to acidic phases using positively charged carrier; **b** from acidic to basic aqueous phases using negatively charged carrier [258]

K^+ ions, and against their concentration gradient, pumped by chemical energy. They also found that the OH^- transport is competing with the carboxylate transport. In recent years, the amino acid extraction has been carried out along these lines.

Five types of amino acid separations can be distinguished:

– extraction with a negatively charged carrier at a low pH,
– extraction with a positively charged carrier at a high pH,
– permeation through a liquid membrane from a low pH to a high pH with a negatively charged carrier,
– permeation through a liquid membrane from a high pH to a low pH with a positively charged carrier, and
– separation by reversed micelles.

Carrier extraction of amino acids. Extraction at a low pH has barely been investigated [259]. The main problem is the high degree of coextraction of inorganic cations. Also, only few investigations are known on amino acid extraction at a high pH [260–268].

The equilibrium of D,L-phenlyalanine between trioctyl-methyl-ammonium chloride (TOMAC, Adogen 464)/xylene as well as N-lauryl-N-trialkyl-methyl-amine (Amberlite LA-2)/xylene and aqueous phases [260], the mass transfer kinetics in a stirred cell [261] in a single droplet [262] and in a laboratory-scale pulsed sieve plate column [263], the separation of D,L-tryptophane from D,L-tyrosine [264–266] and aspartic acid from arginine [267, 268] – both of them by means of TOMAC – were investigated at high pH values.

The extractions of D,L-tryptophane and D,L-tyrosine from their NaOH solution with a quaternary ammonium chloride (TOMAC) in xylene serve to make evident the problems of amino acid extraction [264–266].

The following equilibrium reactions prevail in the amino acid (HS) system in the alkaline (0.01 N NaOH) aqueous phase (a) and the quaternary ammonium chloride (Q^+Cl^-) in the organic phase (o):

1) Dissociation of the amino acid

$$HS_a + H_2O \rightleftharpoons H_3O + S_a^- \tag{4.54}$$

2) Reaction of the amino acid anion with the carrier

$$S_a^- + Q^+Cl_o^- \rightleftharpoons Q^+S_o^- + Cl_a^- \tag{4.55}$$

3) Coextraction of OH^- ion with the carrier

$$OH_a^- + Q^+Cl_o^- \rightleftharpoons Q^+OH_o^- + Cl_a^- \tag{4.56}$$

4) Exchange of coextracted OH^- anions against amino acid anions

$$S_a^- + Q^+Cl_o^- \rightleftharpoons Q^+S_o^- + OH_a^- \tag{4.57}$$

To furnish enough amino acid anions for reaction 2), a pH value must prevail in the aqueous phase which is at least two units above the pK value of the N-terminal proton (tryptophan $pK(NH_2) = 9.39$). In the case of tyrosine, it is necessary to consider the formation of dianions (S_a^{2-}) by the protolysis of a phenolic (ph) proton (tyrosine pK (ph-OH) $= 10, 07$) at pH values higher than 12. According to reactions 1)–4), the pH value of the aqueous phase is reduced as the amino acid and the OH anions are extracted from this phase, which causes a repression of the amino acid protolysis. To avoid this, a large buffer capacity must be used, and the coextraction of the buffer anions must be kept at a much lower level than the coextraction of the OH anions.

Figures 4.21, 4.22 show the isotherm distributions of D, L-tryptophane and tyrosine in a buffer/xylene/TOMAC system as functions of the equilibrium pH value at a constant amino acid concentration (5 mM) and at different carrier concentrations (5, 10, 15 mM).

With an increasing molar ratio of the carrier to the amino acid from 1 to 3, the distribution coefficient of tryptophane increases approximately fourfold, and that of tyrosine roughly twofold.

Above pH 10.5 and at the molar ratio of 3, the distribution coefficient of tryptophane is almost ten times larger than that of tyrosine. This allows their separation by extraction.

The experimentally evaluated equilibrium constants of tryptophane and tyrosine are given in Table 4.25.

In Table 4.26 the extraction equilibrium constants K and isoelectric points PI of various amino acids are compiled with trioctylmethylammonium chloride (TOMAC) in the high pH range. Except for tyrosine, TOMAC reacted with the anion of amino acids and OH$^-$ according to the stoichiometric ratio determined by ionic valency. Tyrosine dissociates into divalent anions in the pH range higher than 13.

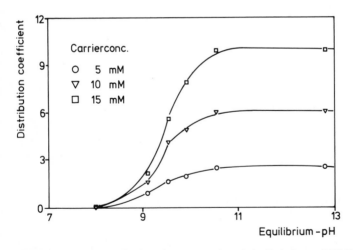

Fig. 4.21. Isotherm distribution of D, L-tryptophane in buffer/xylene + TOMAC. Parameter: carrier concentration. Initial amino acid concentration: 5 mM [264]

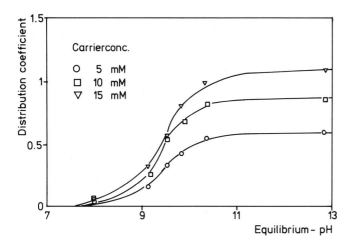

Fig. 4.22. Isotherm distribution of D, L-tyrosine in buffer/xylene + TOMAC. Parameter: Carrier concentration. Initial amino acid concentration: 5 mM [264]

Table 4.25. Experimentally evaluated equilibrium constants of tryptophane and tyrosine [265]

D, L-tryptophan	D, L-tyrosine
$K_2 = \dfrac{[Q^+S_o^-][Cl_a^-]}{[Q^+Cl_o^-][S_o^-]}$ (4.55a)	$K_2 = \dfrac{[Q_2^+So_2^-][Cl_a^-]^2}{[Sa_2^-][Q^+Cl_o^-]^2}$ (4.55b)
$K_2 = 7.0$	$K_2 = 0.056$
$K_3 = \dfrac{[Q^+OH_o^-][Cl_a^-]}{[OH_a^-][Q^+Cl_o^-]}$ (4.56a)	$K_3 = \dfrac{[Q^+OH_o^-][Cl_a^-]}{[OH_a^-][Q^+Cl_o^-]}$ (4.56b)
$K_3 = 4.5$	$K_3 = 4.5$
$K_4 = \dfrac{[Q^+S_o^-][OH_a^-]}{[S_a^-][Q^+OH_o^-]}$ (4.57a)	$K_4 = \dfrac{[Q_2S_o^{2-}][OH_a^-]^2}{[Q^+OH_o^-]^2[S_a^{2-}]}$ (4.57b)
$K_4 = 1.56$	$K_4 = 0.0027$

The equilibrium constant K increases as the number of carbon atoms increases. The highest K value obtained for tryptophan was larger by a factor of about 260 than the lowest for glycine. K values of aromatic amino acids are rather high compared to those of aliphatic amino acids. There is a clear relationship between the K value and the hydrophobicity scale of the amino acids, which is defined as the free energy change when each amino acid transfers from the water to the ethanol phase [274]. With increasing hydrophibicity scale, the K value increases [273].

During the extraction of amino acids with TOMAC at high pH values, OH$^-$ and all types of inorganic anions (SO_4^{2-}, CO_3^{2-}, PO_4^{3-}) are coextracted with comparable extent to hydrophilic amino acids.

Table 4.26. Extraction equilibrium constants K and isoelectric points pI of various amino acids [273]

Amino acid	pI	K
Glycine (Gly)	6.0	0.036
Alanine (Ala)	6.0	0.038
Valine (Leu)	6.0	0.089
Leucine (Val)	6.0	0.29
Isoleucine (Ile)	6.0	0.24
Methonine (Met)	5.7	0.21
Phenylalanine (Phe)	5.5	0.97
Tryptophan (Trp)	5.9	8.89
Tyrosine (Tyr)	5.7	0.40
Histidine (His)	7.6	0.083
Arginine (Arg)	10.8	0.062
Serine (Ser)	5.7	0.049
Threonine (Thr)	6.2	0.071

Similar to the equilibrium, the kinetics of the extraction is also influenced by the coextraction of the OH^- and other anions at a high OH^--ion concentration, which is necessary for the formation of amino acid anions. To evaluate the kinetics of the amino acid anion formation and the OH^- coextraction as a function of the OH^- ion concentration, only the extraction of a single amino acid (tryptophane) from 0.01 N NaOH by TOMAC in xylene has been investigated. It was possible to evaluate the extraction kinetics of both because of the presence of only one single coextractive component (no buffer anions other than OH^-) and by measuring the concentrations of the amino acid (by photometer) and of Cl^- ions (by an ion-selective electrode) in the aqueous phase. Figure 4.23

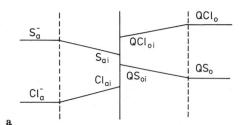

a

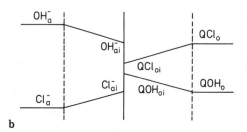

b

Fig. 4.23. Schematic concentration profiles of the components: amino acid anion S_a^- chloride anion Cl_a^- in the bulk of the aqueous phase and S_{ai}^- and Cl_{ai}^- at the interface in this phase, quaternary ammonium chloride $Q^+Cl_{oi}^-$ and amino acid ammonium salt $Q^+S_o^-$ in the organic bulk phase and $Q^+Cl_{oi}^-$ and $Q^+S_{oi}^+$ at the interface of this phase [264]: **a** amino acid extraction; **b** OH^--ion extraction

shows the concentration profiles at the interface for both of these extractions. Since the amino acid is almost completely dissociated, the anion is insoluble in the xylene phase, and the quaternary ammonium chloride in insoluble in the aqueous phase, the reaction occurs at the interface.

These investigations were performed in a stirred cell (for more details, see [265, 266]). From Fig. 4.24, one can detect that proportionally more Cl^- ions are extracted from the extracted amino acid than expected. This is due to the coextraction of OH^- ions. The coextracted OH^- ion flux is calculated from the molar Cl^- and amino acid ion fluxes. In Fig. 4.25, the ratio of the molar fluxes of amino acid and Cl^- ions are plotted as a function of the extraction time. The ratio $\Phi = 1$ corresponds to the absence of coextraction. One can see that Φ is always lower than unity.

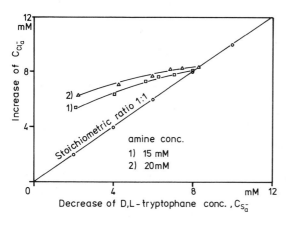

Fig. 4.24. Relationship between the increase of Cl^- ion concentration and the decrease of the D,L-tryptophane concentration in the aqueous phase (0.01 M NaOH) at two different carrier concentrations in the TOMAC/xylene system [264]

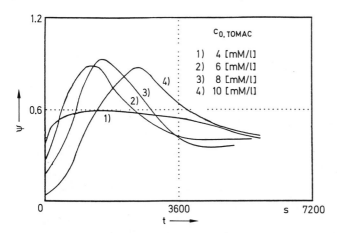

Fig. 4.25. The ratio of the amino acid flux J_S to the chloride flux J_{Cl}, Φ, as a function of the extraction time. Parameter: TOMAC concentation in xylene. Initial concentration of tryptophan = 2 mM [264].

The investigations indicate that the equilibrium of the amino acid extraction is attained after about 1 h, independent of the concentrations of amino acid in the feed and the carrier. After that the OH^- ion concentration increases linearly with time until the carrier is exhausted. However, the Φ-maximum is already attained at about 1/2 h.. This indicates that the extraction of tryptophane is a complex process.

Similar results were obtained with the extraction of aspartic acid and arginine by means of TOMAC in xylene [268]. The third dissociation constant of aspartic acid is at 9.98 and that of arginine at 13.3. Therefore, they can be separated at pH 11.5, where the degree of extraction of aspartic acid has its maximum.

The main problem is the high degree of coextraction of OH^- ions. It can be reduced by diminishing the TOMAC concentration in the organic phase, but it is still too high for an economically sound separation process.

Generally speaking, a separation of amino acids is only possible if their highest (second or third) dissociation constants are far enough apart, i.e., by at least two units. The optimal pH, in this case, is between the two dissociation constants. High pH values are needed for the formation of negatively charged ions because of the high second or third dissociation constants of amino acids. Under these conditions, the coextraction of OH^- ion is very high.

Separation of amino acids by liquid membrane. Figure 4.26 illustrates the principles of the liquid membrane technique. Two miscible aqueous phases are separated by an organic (membrane) phase. The membrane system is produced in two stages. In the first stage the internal aqueous solution is emulsified in the organic membrane phase with low solubility in the aqueous phase at high (10 000 rpm) shear rates. A sufficient stability of the emulsion is attained by using suitable surfactants (Span 80). The stable emulsion contains small droplets 1 to 10 μm in diameter, and is dispersed in the continuous aqueous phase at low stirred speed (200–300 rpm). This yields droplets 1–3 mm in diameter. The permeation of the solute from the outer to the inner aqueous phase across the organic membrane phase is facilitated by carriers soluble only in the organic phase.

Behr and Lehn [258] were also the first ones to report on the separation of amino acids that permeated through a liquid membrane with a highly lipophilic positively charged carrier from an alkaline solution to an acidic solution, or through a liquid membrane with a highly lipophilic negatively charged carrier from the acidic solution to an alkaline solution. They investigated the transport of phenylalanine, tryptophane, leucine, tyrosine, valine, alanine, glycine, serine, and the dipeptides glycyltyrosine, glycylphenylalanine, phenylalanylglycine as well as that of the OH^- ion.

Lysine permeated from an encapsulated acidic solution (1 N HCl) through a vacuum oil membrane with D2EHPA as carrier and sorbitoleate as surfactant into the outer phase (aqueous buffer). However, the authors mainly investigated membrane swelling and stability [256]. Contrary to this, L-phenylalanine was

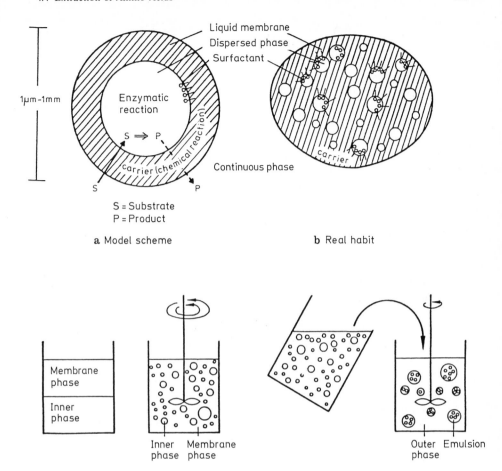

Fig. 4.26a, b. Principles of the liquid membrane technique [272]: **a** structure of the liquid membrane: model scheme (*S*, substrate; *P*, product) (*left*) and real habitat (*right*); **b** formation of liquid membrane: *emulsification* of the aqueous phase in the organic membrane phase at high stirrer speed (*left*) and dispersion of the emulsion in the outer aqueous phase at low stirrer speed (*right*)

extracted from the alkaline (NaOH) solution as exterior phase at pH 11 through a paraffin membrane with 1% Aliquat 336 as carrier, 4% paranox 100 as surfactant, and 5% decyl alcohol as cosurfactant into the alkaline (KOH + 2 M KCl) solution as interior phase at pH 11 [257]. After 40 minutes, the membrane swelling and stability as well as the influence of inorganic anions in the exterior phase, surfactant type and carrier concentration in the membrane phase on the solute concentration in the membrane phase on the solute concentration in the inner phase were investigated again. It became apparent that the permeation rate decreases with increasing sulfate ion concentration and passes a maximum at 3 vol% carrier concentration. The same investigations are discussed in more detail in [269]. An extraction of phenylalanine with D2EHPA

by liquid membrane was reported by Itoh et al.. [277]. Hano et al. [276] found
that the equilibrium constants as well as the extraction rates of different amino
acids are correlated to their hydrophobicity scale.

In contrast to these fundamental investigations, the use of liquid membrane
for the separation of D- and L-amino acids has already been realized on
a laboratory scale in continuous operation [271, 272]. The example of the
separation of L-phenylalanine from D, L-phenyl-alanine methylester by means of
α-chymotrypsin (E.C.3.4.21.1) immobilized in a liquid membrane emulsion
serves to elaborate on the principle of this separation technique in somewhat
more detail. The aqueous enzyme solution is emulsified in the membrane phase
(kerosene, paraffin, or cyclohexane) in presence of Span 80. The permeation of
the substrate (D, L-phenylalanine methyl ester) from the outer aqueous phase (at
25° and pH 6) through the membrane is due to physical solubility. The enzyme
(α-chymotrypsin), encapsulated in the inner phase, reacts only with the L-ester
(Fig. 4.27). The product, L-phenylalanine is enriched in the inner phase and the
unchanged D-ester permeates through the liquid membrane – due to physical
solubility into the outer phase. The reaction was performed in batch as well as in
continuously operated reactors. In batch runs, the emulsion was kept back by
means of a solid membrane mounted on the bottom of the reactor, for (outer
phase) sampling, on-line chemical analysis, and pH control. Figure 4.28 shows
the conversion of the L-ester as a function of the mean residence time of the outer
phase in the reactor and the enzyme activity as parameter in a continuous runs
with an emulsion phase (immobilized biocatalyst) retained by a solid membrane
on the bottom of the reactor.

Amino acid separation by reversed micelles. Reversed micelles are formed by
surfactants in organic solvents. The polar group (heads) of the surfactant
molecules are directed towards the interior of the spheroidal aggregate forming
a polar core, and the aliphatic chains are directed towards the organic solvent.
This is the reverse of the situation in normal micelles in water (Fig. 4.29) [281].
They are mainly used for solubilization of proteins in organic solvents.

Fig. 4.27. Separation of L-phenylalanine from D, L-phenylalanine methyl ester by enzymatic conversion [272]

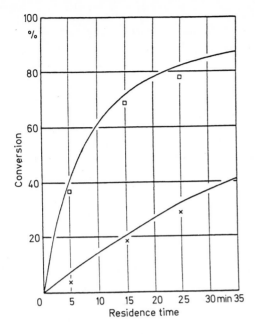

Fig. **4.28.** Conversion of L-phenylalanine methylester as a function of the mean residence time of the continuous phase in the reactor at different enzyme concentrations: □ 70.0 mg, × 5.25 mg enzyme [272]

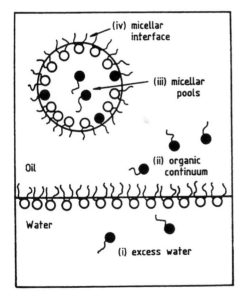

Fig. **4.29.** Schematic picture of a reversed micelle with four solubilization environments in two-phase system at equilibrium [278]

The extraction of amino acids with reversed micelles was investigated by Hatton and coworkers [278, 279]. They used Aerosol-OT (bis(2-ethylhexyl) sodium sulfo-succinate) (AOT) as surfactant in isooctane as well as dodecyl-trimethylammonium chlorine (DTAC) as surfactant in heptane/hexanol and contacted these organic phases with amino acid in phosphate buffer solution at pH 6 to 6.5. After shaking these phases and phase separation, the upper organic phase contained the main part of the amino acids in reversed micellar solution. Leonidis et al. [278, 279] evaluated the partition coefficient of the amino acids between the aqueous and organic phase. They also considered the physico-chemical basis of this phenomenon and found that the free energy of transfer of a large number of amino acids from water to the surfactant interfaces of AOT/isooctane and DTAC/heptane/hexanol W/O microemulsions correlate well with the existing hydrophobicity scales [279]. It is possible to calculate the partition coefficient P_c^S of the amino acids between water and the surfactant interface of AOT and DTAC-reversed micelles by using simple phase-equilibrium experiments:

$$P_c^S = \frac{55.5 V_{a,in}}{N_{S,tot}(1 + R_C)} \left[\frac{C_{a,in} - C_{a,f}}{C_{A,f}} \right] \tag{4.58}$$

where $V_{a,in}$ is the initial volume of the aqueous phase before the contact, $N_{S,tot}$ the total moles of the surfactant, R_C the ratio of moles of alcohol (octanol, heptanol) to moles of surfactant (AOT, DTAC) in the reversed micellar interface: 2.88, and $C_{a,in}$ and $C_{a,f}$ are the molar concentrations of the solute (amino acid) in aqueous phase before the contact (in) and for the final aqueous phase (f).

In Table 4.27, some partition coefficients of amino acids are compiled.

Furusaki and Kishi [280] pointed out, based on the investigations with Tyrosine and Arginine and AOT in reversed micelles, that amino acids with different isoelectric points could be separated by this technique The distribution equilibrium was, however, only slightly sensitive to the pH value changes.

Adachi et al. [282] extracted amino acids with different properties with AOT in reversed micelles and evaluated their partition and equilibrium coefficients between the organic and aqueous phases. They found that hydrophilic amino acids are entrapped in the core water of the microemulsion globule, whereas amino acids with hydrophobic groups are mainly incorporated in the interface of the globule.

Table 4.27. Interfacial partition coefficients of some amino acids [279]

Compound	P_c^S (water → AOT)
Valine	4.6
Leucine	26.1
Isoleucine	17.0
Proline	2.2
Phenylglycine	16.8
Phenylalanine	89.3
Tyrosine	9.0
Tryptophan	238.1
Methionine	19.2

4.4.1 Symbols for Sect. 4.4

Cl_a^-	chloride anion in the aqueous phase
HS_a	amino acid in the aqueous phase
K_2, K_3, K_4	equilibrium constants (Table 4.25)
OH_a^-	OH^- anion in the aqueous phase
$Q^+Cl_o^-$	quaternary ammonium chloride in the organic phase
$Q^+OH_o^-$	quaternary ammonium hydroxide in the organic phase
$Q^+S_o^-$	quaternary ammonium salt of the amino acid in the organic phase
S_a^-	amino acid anion in the aqueous phase

4.5 Recovery of Antibiotics from Fermentation Broths

An antibiotic is "a chemical substance derived from a microorganism which has the capacity of inhibiting growth and even destroying other microorganisms in dilute solutions" – as defined by Selman Wachsman in 1942. Later this definition was extended to all microbial compounds which are able to selectively affect various biochemical growth processes at minimal concentration in humans, animals, plants, or microorganisms [284]. This definition covers a large group of compounds of entirely different chemical structures and character.

Most antibiotics are excreted by the cells into the broth. The cells are removed from the broth by filter press or a rotary filter (often with filter aids) and the product is recovered from the broth usually by extraction or by absorption [283]. Also in the case of adsorption, the desorption from the sorbent is an extraction process. Sometimes the product is partly intracellular. In this case, the cells are removed from the broth again, but the solid (sometimes filter-aid-containing) cake is also extracted. The broth and the cells are seldom extracted together.

The extract is usually filtered and purified further (by reextraction or by adsorption). If the product concentration is high enough after the reextraction by an aqueous solution, a precipitation or crystallisation of the product is possible.

If the product is a weak acid with a low pK_a value, the pH of the broth has to be reduced considerably below the pK_a value to be able to extract soluble-free acid from the organic solvent.

If the product is a weak base with a high pK_a value, the pH of the broth has to be increased above the pK_a value by two units to be able to extract soluble-free base from organic solvent.

If the product is very soluble in water, the broth must be saturated by salt (salting out) to increase the degree of extraction with the organic solvent.

Carbon-bonded, oxygen-bearing extractants (alcohols, esters and ketones) are often used for the recovery of polar compounds. For the extraction of hydrophobic compounds, apolar extractants (petrolether, methylene dichloride, ethylene dichloride) are applied. The main problem with the carbon-bonded, oxygen-bearing extractants is their relatively high solubility in broth. With increasing molecular weight, their solubility in the aqueous phase decreases, but, their tendency to form stable emulsion increases.

In Table 4.28, several examples of solvent extraction of antibiotics are compiled. More details are given in the excellent book of Vandamme [284], and in the excellent handbook of Atkinson and Mavituna [283].

In the Handbook of Solvent Extraction of Lo, Baird, and Hanson, [302], only a short chapter considers solvent extraction in pharmaceutical manufacturing processes.

Figure 4.30 shows the flow sheet of the production, recovery, and purification process of bacitracin, which uses a filter press [283], and Fig. 4.31 depicts the flow sheet of the penicillin purification process of Gist-Brocades, which uses a rotary filter for the extraction process [284]. In both processes, the extract is clarified by charcoal treatment. Because of the high product concentrations in the broth after these recovery and purification steps, a product crystallization is possible.

Except for penicillin, which is treated in Chapter 4.6, only few extraction processes for antibiotics were considered in detail in the literature. Earlier publications on tetracycline and cephalosporin extractions do not have practical relevance any longer, since nowadays tetracycline is recovered by precipitation and cephalosporin by adsorption.

Table 4.28. Solvent extraction of some antibiotics

Product	Medium	Solvent	pH	Ref.
actinomycin	cake	1 MeOH + 2MCl$_2$	2.5	[285]
adrianimycin	cake	acetone	acid.	[286]
bacitracin	broth	n-BuoH	7.0	[287]
chloramphenicol	broth	EtAc	N-alk	[288]
clavulanic acid	broth	n-BuOH	2.0	[289]
cyloheximid	broth	MCl$_2$	3.5–5.5	[290]
erythromycin	broth	AmAc	alk.	[291]
fusidic acid	broth	MIBK	6.8	[292]
griseofulvin	cake	BuAc, MCl$_2$	n	[293]
macrolides (general)	broth	MIBK, EtAc	alk	[294]
nisin	broth	CHCl$_3$ + sec-Oct.alc.	4.5	[295]
	mash	CHCl$_3$	2.0	[295]
oxytetracycline	broth	BuOH		[296]
penicillin G	broth	BuAc, AmAc	2.0	[297]
salinomycin	cake	BuAc	9.0	[298]
tetracycline	broth	BuOH		[299]
tylosin	broth	AmAc, EtAc,		[300]
virginiamycin	broth	MIBK	acid.	[301]

MeOH (methanol), AmAc (amylacetate), BuOH (1-butanol), BuAc (butylacetate) EtAc (ethylacetate), MCl$_2$ (methylene dichloride), MIBK (methylisobutylketone), CHCl$_3$ (chloroform), n (neutral) acid. (acidic), alk. (alkaline), broth (cell-free medium), cake (medium-free cells), mash (broth with cells)

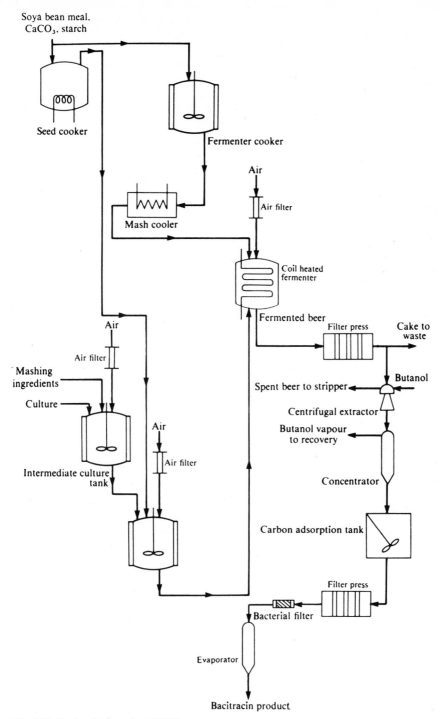

Fig. 4.30. Bacitracin flow sheet [283]

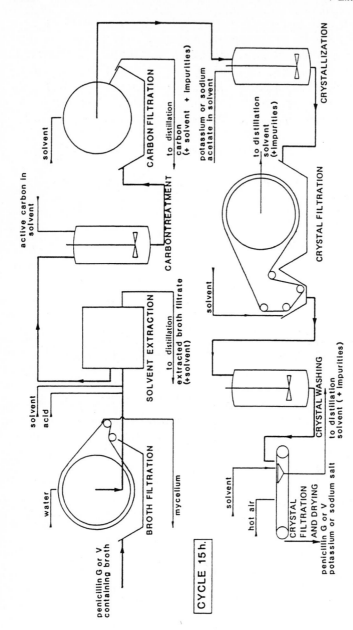

Fig. 4.31. Penicillin purification process of Gist-Brocades [297]

A Russian research group investigated the extractive recovery of erythromycin thoroughly [303–306]. They found that the nature of the extractant (butyl acetate, amyl acetate, ethylene dichloride, chloroform and methylene dichloride) has no effect on the purity of the final product, but that the acetate esters have the highest degree of extraction. Butyl acetate in combination with organic (lactic, citric, malic and tartaric) acids were found to yield a higher degree of extraction than butyl acetate alone. The highest yield (82%) was attained with lactic acid. The extraction with butyl acetate is influenced by the pH and the temperature. The optimal conditions were evaluated and published [306].

The product quality can be improved by the addition of potassium sulfate to the buffered aqueous solution, transferring the separated precipitate into butyl acetate, and bringing the pH of the erythromycin solution to 6.5 [302, 305].

The recovery of several antibiotics (e.g., penicillin, erythromycin) by extraction is usually performed by centrifugal extractors (e..g., Podbielniak, Alfa-Laval, Westfalia).

4.6 Recovery of Penicillin from Fermentation Broths

The recovery of penicillin G is the best investigated antibiotic extraction process. Therefore, this process will be considered in detail. As already pointed out, the mycelium is usually separated by a rotary vacuum filter and is washed on the filter.

With an overlong fermentation time when the protein content in the broth increases by mycelial lysis, a sharp decrease in filter-ability occurs. Also the phase separation after the extraction is impaired by the high protein content.

By means of the extraction decanter (see Fig. 3.5, Chap 3) the entire broth extraction is also carried out.

The mycelium-free broth is cooled down to low temperatures (0–3 °C), acidified with sulfuric acid (pH 2.0–3.0) and usually extracted by amyl acetate or butyl acetate in centrifugal extractors. In order to repress the emulsion forma-

Table 4.29. Distribution coefficients for penicillin G at pH 4 between aqueous solutions and the listed solvents [312 and cited in 311]

Methyl cyclohexanone	180
Dimethyl cyclohexanone	160
Methyl cyclohexanol	80
2-Chloro-2^1-methoxyl diethyl ether	57
Cyclohexyl acetate	62
Furfuryl acetate	44
Methyl isobutyl ketone	33
Dimethyl phthalate	30
2-Ethyl hexanol	26
Amyl acetate	20
Diethyl oxalate	20

tion, a deemulsifier is added to the broth. A typical 3:1 or 4:1 broth-to solvent ratio is used. This organic solution is contacted with an aqueous buffer solution at pH 6 to produce a penicillin-rich aqueous solution. After reacidification, it is extracted again with butyl acetate or amylacetate. [307–310].

The first distribution coefficient data were published by Rowley et al. [312] (Table 4.29). The variation of the partition coefficient with the pH was investigated by Kansava et al. [313]. Different centrifugal extractors were tested (Luwesta [314], Podbielniak [315] and Alfa-Laval [316]). However, no quantitative comparisons have been published yet.

From Table 4.29 it is obvious that with several extractants higher distribution coefficients can be attained than with amyl acetate. However, partition coefficients that are too high, cause problems during reextraction.

A new development is the direct extraction of the mycel containing broth with a countercurrent extraction decanter [317–319] (Fig. 3.5, Chap 3). 90% degrees of extraction have been attained in a single-stage unit, and 95% degrees in a two-stage unit. Several companies have adopted this technique recently and use it for commercial production [340].

In spite of the commercial production of penicillins, very little quantitative information became known about their extraction. Therefore, systematic investigations were carried out with penicillin G – which is the bulk product – in the research group of Schügerl [320–336] to improve the recovery process. The following paragraphs give the result of these investigations.

Stability of Penicillin G [320, 337]. All penicillins have a common basic structure (a condensed thiazolidine-β-lactam ring system),

where * stands for asymmetric carbon atoms and X^+ for cations. Penicillin G (benzyl-penicillin) with

and Penicillin V (phenoxymethylpenicillin) with

were used for the investigations Their properties are given in Table 4.30.

Penicillin G is a weak acid ($pK_a = 2.75$). The same is true for penicillin V ($pK_a = 2.70$). Only the free undissociated acid can be extracted by organic

Table 4.30. Properties of free penicillins and their potassium salts [344]

Compound	Molecular weight	Activity $IU^a mg^{-1}$	Decom. point °C	Solubil. in water
Penicillin G, H^+	334.4	1777		slightly sol.
Penicillin G, K^+	372.2	2598	214–217	well
Penicillin V, H^+	350.4	1699	120–128	well
Penicillin V, K^+	388.5	1532	256–260	well

[a] IU, international unit

extractants. Therefore, the extraction must be performed below the pK_a value, at pH 2.0–2.5. However, the free acids are unstable at these low pH values (Fig. 4.32). From Fig. 4.32 it is obvious that considerable losses of penicillin G are expected during its recovery by extraction with carbon-bonded, oxygen bearing extractants.

However, in the pH range 5 to 8, in which penicillin is stable, it prevails in the aqueous phase as pencillin acid anion (P^-). A patent of Beecham has pointed out that the solubility of penicillin can be increased by using a phase transfer catalyst, a quaternary ammonium salt (Q^+X^-), to increase the solubility of penicillin in an apolar extractant (methylene dichloride) by forming a salt (Q^+P^-) from the sodium salt of penicillin G (Na^+P^-) and from the ammonium chloride (Q^+Cl^-) [345]. This principle has already been used for the recovery of organic acids (Chapters 4.2–4.4.).

Therefore, systematic investigations were performed with different primary, secondary, tertiary, and quaternary amines (with long hydrophobic groups to keep their solubility in the fermentation broth low) with regard to their use as suitable extractants. Of the large number of investigated amino compounds, 52 are listed with their properties in [321, 328].

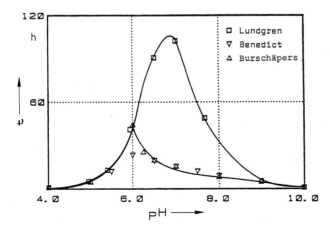

Fig. 4.32. Decomposition times for penicillin G with 5% loss of activity [320]

The stability of pencillin G in aqueous solutions, which were in equilibrium with different amine extractants (in n-butyl acetate), were investigated at pH 5, and it was found that Amberlite LA-2 (which is a mixture of secondary amines of the type N-lauryl-N-trialkylmethylamine $C_{12}H_{25}$-NH-CR$_1$R$_2$R$_3$ with $R_1 + R_2 + R_3$ equal to 11-13 carbon atoms) and Adogen 283 (tris(tridecyl)amine) have a stabilizing effect. The others are neutral or have destabilizing effect on penicillin G. At pH 7, LA-2 has a strong and Amberlite 283 a slight stabilizing effect. Generally speaking, the decomposition rate of penicillin G in n-butyl acetate in the presence of a carrier is by one order of magnitude less than in the aqueous phase.

Distribution Coefficients and Equilibrium Degrees of Extraction [320, 321, 337]. In the absence of amine extractants, the distribution coefficient K_p (p means physical extraction with carbon-bonded, oxygen-bearing extractants without chemical reaction) of the undissociated penicillin acid (HP), which is soluble in the aqueous (a) and organic (o) phases, is given by

$$K_p = \frac{C_{HPo}}{C_{HPa} + C_{Pa}}, \tag{4.59a}$$

where C_{Pa} is the concentration of the penicillin acid anion in the aqueous phase. K_p depends on the pH value:

$$K_p = P_c \frac{1}{1 + 10^{pH - pKa}}, \tag{4.59b}$$

where

$$P_c = \frac{C_{HPo}}{C_{HPa}} \tag{4.60}$$

is the partition coefficient of the free acid on a molar base.

The degree E of extraction is defined as the fraction of the extracted penicillin with regard to its overall concentration

$$E = \frac{C_{HPo}}{C_{HPo} + C_{HPa} + C_{Pa}}, \tag{4.61a}$$

$$E = \frac{100}{1 + (1 + 10^{pH - pKa})/P_c}. \tag{4.61b}$$

The reaction of the acid with a primary, secondary, or tertiary amine extractant corresponds to a neutralization reaction, Amine A, dissolved in the organic phase reacts with the penicillin acid anion P^- and proton H^+ in the aqueous phase:

$$A_o + P_a^- + H_a^+ \rightleftharpoons AHP_o \tag{4.62}$$

Reaction (4.62) is an instantaneous reaction; thus the extraction rate is controlled by the mass transfer process The equilibrium constant K_G of reaction

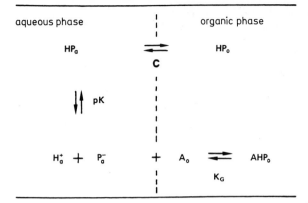

Fig. 4.33. Scheme of extraction of penicillin with aliphatic secondary or tertiary amines [320]

(4.62) is given by

$$K_G = \frac{C_{AHP}}{C_A C_P C_H},$$ (4.63)

where the charge is omitted for simplicity's sake.

The following relationship holds true for the penicillin anion concentration, because of the simultaneous physical and chemical (with amine carrier) extractions (Fig. 4.33) [320]:

$$C_p = -0.5\left[\frac{C_{At} - C_{Pt}}{Z} + \frac{1}{K_G C_H}\right] + \left\{0.25\left[\frac{C_{At} - C_{Pt}}{Z} + \frac{1}{K_G C_H}\right]^2\right.$$
$$\left. + \frac{C_{Pt} C_P}{K_G C_H Z}\right\}^{0.5},$$ (4.64)

with

$$Z = 1 + 10^{pKa - pH} + P_c 10^{pKa - pH},$$ (4.64a)

$$C_{Pt} = C_P + C_{HPa} + C_{HPo} + C_{AHP},$$ (4.65)

the overall concentration of pencillin, and

$$C_{At} = C_A + C_{AHP},$$ (4.66)

the overall concentration of amine.

From C_P we obtain the relationship for the distribution coefficient K:

$$K = \frac{C_{Pt}}{C_P(1 + 10^{pKa - pH})} - 1$$ (4.67)

and the degree of extraction

$$E = \left\{1 - \frac{C_P(1 + 10^{pKa - pH})}{C_{Pt}}\right\} 100.$$ (4.68)

The special case of non-stoichiometric extraction is also treated in [321]. In case of penicillin extraction with primary amines in n-butyl acetate, super-stoichiometric extraction ($A_n(HP)$ with $n > 1$) was observed. The identified n values are published in [337].

Since several other acid anions are present (phenyl acetic acid as precursor, penilloic and penicilloic acids as decomposition products, and e.g., citric acid as buffer) in real fermentation broths, their coextraction must be considered as well. Coextraction consumes amine extractant·; therefore, it must be taken into account. The coextraction of anion X^- is described by

$$A_o + X_a^- + H_a^- \rightleftharpoons AHX_o. \tag{4.69}$$

Reaction (4.69) has the equilibrium constant

$$K_X = \frac{C_{AHX}}{C_A C_H C_X}. \tag{4.70}$$

The amine balance is given by

$$C_{At} = C_A + C_{AHP} + C_{AHX}. \tag{4.66a}$$

For the calculation of K_G it is necessary to know K_X:

$$K_G = C_{AHP} \left\{ C_H \frac{C_P + C_{HPa}}{1 + 10^{pKa - pH}} (C_t - C_{AHP} - C_{AHX}) \right\}^{-1} \tag{4.67a}$$

where

$$C_{AHX} = -0.5 \left(C_{AHP} - C_{At} - C_{Xt} - \frac{1}{K_X C_H} \right)^{-1}$$
$$- \left\{ 0.25 \left(C_{AHP} - C_{At} - C_{Xt} - \frac{1}{K_X C_H} \right)^2 - C_{Xt}(C_{At} - C_{AHP}) \right\}^{0.5} \tag{4.71}$$

and

$$C_{AHP} = C_{Pt} - (C_P + C_{HPa}) \left[1 + \frac{P_c}{1 + 10^{pH - pKa}} \right]. \tag{4.72}$$

If the physical coextraction of the weak acid has to be taken into account, these equations must be modified. The reaction of the acid with a quaternary ammonium salt (Q^+Cl^-) corresponds to an ion exchange:

$$Q^+Cl_o^- + P_a^- \rightleftharpoons Q^+P_o^- + Cl_a^- \tag{4.73}$$

The equilibrium constant of the reaction (4.73) is given by

$$K_G = \frac{C_{QP} C_{Cl}}{C_{QCl} C_P} \tag{4.74}$$

since $C_{QP} = C_{Cl}$, it follows that

$$K_G = \frac{(C_{Pt} - C_P)^2}{C_{QClt}C_P - C_{Pt}C_P + C_P^2}.$$ (4.75)

In the neutral pH range, the reaction (4.73) is independent of the pH value. In the acidic range, the dissociation of penicillin must be taken into account. This pH independence has disadvantages, since at high equilibrium constants the reextraction is difficult.

The equilibrium concentration of C_P is given by

$$C_P = -\frac{K_G C_{QClt} - K_G C_{Pt} + 2C_{Pt}}{2K_G - 1)} + \left\{ \left[\frac{K_G C_{QClt} - K_G C_{Pt} + 2C_{Pt}}{2(K_G - 1)} \right]^2 \right.$$
$$\left. + \frac{C_{Pt}^2}{K_G - 1} \right\}^{0.5}$$ (4.76)

For the coextracted anions similar relationships are valid.

The experimental determination of the distribution coefficients were performed with separatory funnels at 20 °C at different pH values.

In Fig. 4.34a, the degree of extractions are shown as a function of the pH value with different carbon-bonded, oxygen-bearing extractants and chloroform. In Fig. 4.34b, the logarithm of the equilibrium constants K are shown for physical extraction as a function of the pH value.

According to Fig. 4.34a, the highest degrees of extraction are attained with n- and iso-butyl acetate, amyl acetate and chloroform. However, very low pH values are necessary to achieve high degrees of extraction E. In this low pH range, penicillin is very unstable. This is the reason why the recovery losses are considerable, in spite of the low extraction temperatures. When using different amounts of LA-2 as extractant in n-butyl acetate at the same pH values, the degrees of extraction increase with increasing amine amount. At constant amine concentration, E increases with decreasing pH. However, at moderate pH values (pH 5–6) considerable E values can be attained (Fig. 4.35). E.g., at pH 5 and with an amine to penicillin ratio of two, high degrees of extraction (above 90%) can be achieved in comparison with E (about 20%) attainable in the absence of LA-2. The same data are plotted as log K vs pH in Fig. 4.36. One can recognize the nearly parallel shift of log K with increasing amine concentration. In Figs. 4.34–4.36 the symbols are measured and the curves are calculated values. One can recognize that the theoretical relationships developed for the calculations of the partition coefficients and the equilibrium constants are suitable to describe the variation of P_C, E and K_G values as functions of the pH. By means of these measurements the P_c and K_G values were identified. In Table 4.31, partition coefficients and equilibrium constants are given for penicillin G with several solvents and different amine extractants in n-butyl acetate. Also, in the following figures, the symbols indicate measured and curves indicate calculated data.

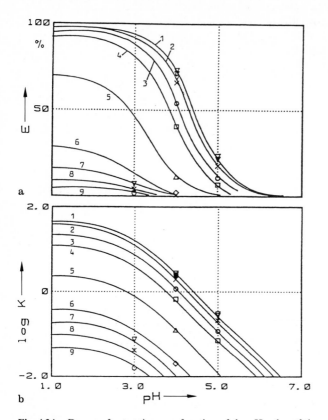

Fig. 4.34a. Degree of extraction as a function of the pH value of the medium for the extraction of penicillin with various solvents [321]: *curve 1,* ▽, *n*-butyl acetate, $P_c = 48$; *curve 2,* ×, isobutyl acetate, $P_c = 37$; *curve 3,* ○, isoamyl acetate, $P_c = 22$; *curve 4,* □, chloroform, $P_c = 12.5$; *curve 5,* △, disiopropyl ether, $P_c = 2.4$; *curve 6,* ◇, dioctyl ether, $P_c = 0.4$; *curve 7,* ▽, *n*-hexane, $P_c = 0.2$; *curve 8,* ×, xylene, $P_c = 0.1$; *curve 9,* ○, kerosene, $P_c = 0.05$. **b** Extraction of penicillin G with different solvents [337]. Logarithm of the distribution coefficients as a function of the pH value: *curve 1,* ▽, *n*-butylacetate; *curve 2,* ×, iso-butyl acetate; *curve 3,* ○, iso-amyl acetate; *curve 4,* □, chloroform; *curve 5,* △, diisopropylether; *curve 6,* ◇, dioctylether; *curve 7,* ▽, *n*-hexane; *curve 8,* ×, xylene; *curve 9,* ○, kerosine

One can see from Fig. 4.35 that the reextraction with an increasing pH is favoured. At pH 8, a high degree of reextraction can be attained. The advantage of the use of primary, secondary, and tertiary amines is an easy way to control the extraction and reextraction. This is in contrast to the use of quaternary amines. As will be discussed later, the reextraction of penicillin from the quaternary ammonium salt is a difficult problem.

In Fig. 4.37, E is plotted as a function of pH value with different amounts of tertiary amines in *n*-butyl acetate. Again, with amine excess, high E values can be attained at pH 5.

In Fig. 4.38, the degrees of extraction are plotted as a function of the pH value for the extraction with LA-2 in different solvents. The highest E values are

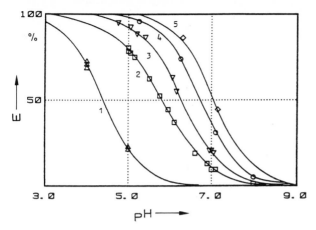

Fig. 4.35. Degree of extraction as a function of the pH value of the medium for the extraction of penicillin G with Amberlite LA-2 and *n*-butyl acetate, $C_P = 10$ mM [321]: *curve 1*, $\triangle$, $C_A = 0$; *curve 2*, $\square$, $C_A = 10$ mM; *curve 3*, $\triangledown$, $C_A = 20$ mM; *curve 4*, $\bigcirc$, $C_A = 50$ mM; curve 5, $\diamond$, $C_A = 100$ mM

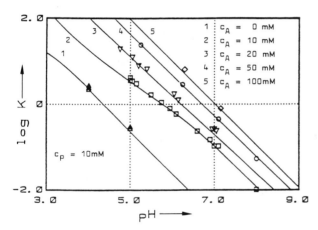

Fig. 4.36. Logarithm of the distribution coefficients as a function of the pH value of the medium [321]. (For symbols, see Fig. 4.35)

attained in iso-amyl acetate, iso-butyl acetate, *n*-butyl acetate and chloroform, which are the best solvents.

In Fig. 4.39, E values are plotted for penicillin V with LA-2 in butyl acetate as a function of the pH value. Already with a stoichiometric amount of amine, high E values are attained at pH 5.

Should a breakdown occur in the later phase of the penicillin fermentation, it can happen that the precursor is present in the broth and the extraction of penicillin still remains economical. Therefore, the E values of the precursors for penicillin G (phenyl acetic acid) and penicillin V (phenoxy acetic acid) were also evaluated with LA-2 in butyl acetate and plotted in Figs. 4.40, 4.41. The

Table 4.31. Partition coefficients P_c and equilibrium constant K_G ($l^2 \, mol^{-2}$) several penicillin G-, and precursor-amine systems in different solvents [337]

Amine	Solvent	P_c	K_G ($l^2 \, mol^{-2}$)
Penicillin G			
—	n-butylacetate	48.0	—
—	iso-butylacetate	37.0	—
—	iso-amylacetate	22.0	—
—	chloroform	12.5	—
—	diisopropylether	2.4	—
—	n-hexane	0.2	—
—	dioctylether	0.4	—
—	xylene	0.1	—
—	kerosin	0.05	—
Amberlite LA-2	n-butylacetate	48.0	1.25×10^8
Dioctylamine	n-butylacetate	48.0	7.0×10^8
Adogen 283	n-butylacetate	48.0	7.5×10^8
Amberlite LA-1	n-butylacetate	48.0	4.0×10^7
Trihexylamine	n-butylacetate	48.0	2.6×10^7
Trioctylamine	n-butylacetate	48.0	2.8×10^7
Tridecylamine	n-butylacetate	48.0	3.1×10^7
Tridodecylamine	n-butylacetate	48.0	3.7×10^7
Alamin 336	n-butylacetate	48.0	2.8×10^7
Adogen 383	n-butylacetate	48.0	2.0×10^7
Adogen 364	n-butylacetate	48.0	2.2×10^7
Adogen 381	n-butylacetate	48.0	2.2×10^7
Dimethyl-myristylamine	n-butylacetate	48.0	4.5×10^7
Dimethyl-Palmithylamine	n-butylacetate	48.0	4.5×10^7
Amberlite LA-2	iso-butylacetate	37.0	1.0×10^8
Dioctylamine	iso-butylacetate	37.0	7.0×10^8
Tryoctylamine	iso-butylacetate	37.0	2.2×10^7
Tridodecylamine	iso-butylacetate	37.0	2.2×10^7
Amberlite LA-3	n-butylacetate[1]	48.0	5.0×10^8
n-Octylamine	n-butylacetate[2]	48.0	3.0×10^{11}
n-Decylamine	n-butylacetate[3]	48.0	1.4×10^{11}
n-Dodecylamine	n-butylacetate[4]	48.0	2.5×10^{11}
Tridecylamine	n-butylacetate[5]	48.0	5.0×10^{11}
Adogen 115-D	n-butylacetate[6]	48.0	2.5×10^{10}
Amberlite LA-2	iso-amylacetate	22.0	8.0×10^7
Amberlite LA-2	diisopropylethane	2.4	1.0×10^6
Amberlite LA-2	dioctylether	0.4	1.24×10^6
Amberlite LA-2	xylene	0.1	4.0×10^6
Amberlite LA-2	kerosin	0.05	2.5×10^5
Amberlite LA-2	chloroform	12.5	2.75×10^9
Dioctylamine	chloroform	12.5	7.0×10^9
Adogen 283	chloroform	12.5	3.0×10^9
Penicillin V			
Amberlite LA-2	n-butylacetate	210.0	3.75×10^8
Dioctylamine	n-butylacetate	210.0	3.0×10^9
Phenylacetic acid			
Amberlite LA-2	n-butylacetate	25.0	2.0×10^7
Phenoxyacetic acid			
Amberlite LA-2	n-butylacetate	30.0	5.0×10^7

[1] m = 1.2; [2] m = 3.0; [3] m = 2.3; [4] m = 2.7; [5] m = 1.5; [6] m = 1.4.

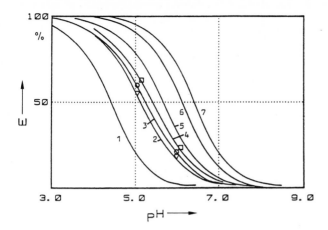

Fig. 4.37. Degree of extraction as a function of the pH value for the extraction of penicillin G with tertiary amine and *n*-butyl acetate, $C_P = 10$ mM [321]: *curve 1*, no carrier; *curve 2*, ∇, Adogen 383, $C_A = 10$ mM; *curve 3*, $\bigcirc$, trioctylamine, $C_A = 10$ mM; *curve 4*, $\square$, dimethylpalmitylamine, $C_A = 10$ mM; *curve 5*, trioctylamine, $C_A = 20$ mM; *curve 6*, trioctylamine, $C_A = 50$ mM; *curve 7*, trioctylamine, $C_A = 100$ mM

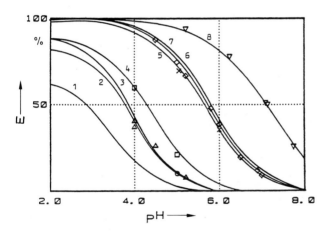

Fig. 4.38. Degree of extraction as a function of the pH value of the medium for the extraction of penicillin G with Amberlite LA-2 in various solvents, $C_P = 10$ mM, $C_A = 10$ mM [321]: *curve 1*, kerosene; *curve 2*, $\triangle$, dioctyl ether; *curve 3*, $\bigcirc$, diisopropyl ether; *curve4*, $\square$, xylene; *curve 5*, $\times$, isoamyl acetate; *curve 6*, $\diamond$, isobutyl acetate; *curve 7*, —, *n*-butyl acetate; *curve 8*, ∇, chloroform

precursors are already extracted fairly well in the absence of LA-2. With increasing LA-2 concentration, the degree of extraction of the penicillins increases more than that of the precursors. Therefore, at high amine concentrations, they are extracted with about the same extraction degree as the pencillins.

Different amines in *n*-butyl acetate were used as extractants for pencillin G and V and compared. Primary amines are not very suitable extractants

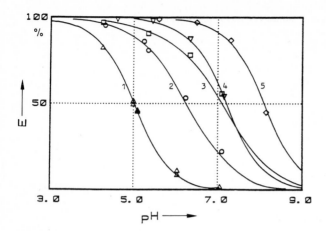

Fig. 4.39. Degree of extraction as a function of the pH value of the medium for the extraction of penicillin V with secondary amines and *n*-butyl acetate, C_P = 10 mM [321]: *curve 1,* $\triangle$, no carrier; *curve 2,* $\bigcirc$, Amberlite LA-2, C_A = 10 mM; *curve 3,* $\square$, dioctylamine, C_A = 10 mM; *curve 4,* $\triangledown$, Amberlite LA-2, C_A = 50 mM; *curve 5,* $\diamond$, dioctylamine, C_A = 50 mM

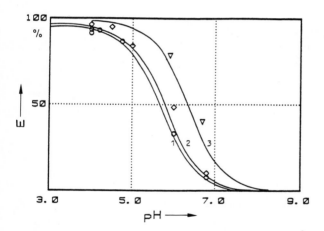

Fig. 4.40. Degree of extraction as a function of the pH value of the medium for the extraction of phenylacetic acid with Amberlite LA-2 and *n*-butyl acetate C_P = 10 mM [321]: *curve 1,* $\bigcirc$, C_A = 0; *curve 2,* $\diamond$, C_A = 10 mM; *curve 3,* $\triangledown$, C_A = 100 mM

because of the moderate partition coefficients of their penicillin salts. In addition, they have a fairly high solubility in water and high interfacial activity, which causes stable emulsions. The highest degrees of extraction were attained by using secondary amines as extractants. When secondary amines are used, penicillin G can be extracted up to a pH value of 6. The distribution coefficients K of their amine complexes with penicillin G markedly depend on the amine structure. A disadvantage of the secondary amines can be their reactivity with solvents, e.g., they may form amides with acetic acid esters during distillative

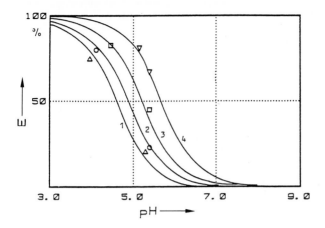

Fig. 4.41. Degree of extraction as a function of the pH value of the medium for the extraction of phenoxyacetic acid with Amberlite LA-2 and *n*-butyl acetate, $C_P = 1$ mM [297]: *curve 1*, $\triangle$, $C_A = 0$; *curve 2*, $\bigcirc$, $C_A = 1$ mM; *curve 3*, $\square$, $C_A = 3$ mM; *curve 4*, ∇, $C_A = 10$ mM

purification of the solvent. (However, this is not necessary, since they can be sufficiently purfied by alkaline solutions).

Tertiary amines are suitable as extractants, but the distribution coefficients of their complexes are significantly lower than those of secondary amine complexes. Their advantage is due to their low reactivity with solvents.

Quaternary amines can also be used for the extraction of penicillin. However, reextraction is difficult, and very large amounts of anions (e.g., Cl^-) are needed to recover the penicillin.

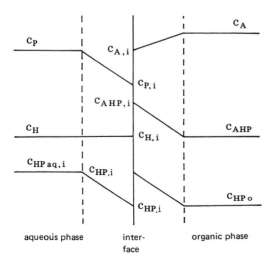

Fig. 4.42. Diagrammatic representation of the concentration profiles during the extraction of penicillin [322]

Penicillin V can be recovered more easily by physical as well as by amine extraction than can penicillin G. Precursors of penicillin G and V can also be coextracted with amines, and the selectivity of penicillin can be improved by using amines as extractants.

The extraction equilibrium of secondary and tertiary amines can be described by a simple stoichiometric reaction. When primary amines are used, the non-stoichiometric reaction must be taken into account. The theoretical relationships developed for the distribution coefficients and degrees of extraction yield useful results which agree well with the measurements.

Kinetics of Extraction [322, 337] The two-film model was used to describe the mass transfer across the liquid interface (Fig. 4.42). During extraction with *n*-butyl acetate or amyl acetate, only the free penicillin acid is extracted, and for the transfer across the interface the following balances hold:

$$j_{HPa} = k_{HPa}(C_{HPa} - C_{HPai}) \tag{4.77a}$$

in the aqueous phase and

$$j_{HPo} = k_{HPo}(C_{HPoi} - C_{HPo}) \tag{4.77b}$$

in the organic phase.

Flux j is controlled by mass transfer coefficients k_{HPa} and k_{HPo}, and the driving forces, i.e., the difference between solute concentrations C_{HPa} in the bulk and C_{HPi} at the interface for the aqueous phase and the difference between solute concentrations C_{HPoi} at the interface and C_{HP} in the bulk for the organic solvent phase.

For the steady state ($j_{HPa} = j_{HPo}$), the interfacial concentrations can be eliminated, this yields

$$C_{HPai} = \frac{k_{HPa}C_{HPa} + k_{HPo}C_{HPo}}{P_c k_{HPo} + k_{HPa}}, \tag{4.78}$$

where $P_c = C_{HPoi}/C_{HPai}$ is the partition coefficient.

For the extraction of penicillin from the aqueous phase with amine in *n*-butyl acetate following balances hold:

$$j_{Pa} = k_P(C_{Pa} - C_{Pai}) \tag{4.77c}$$

for penicillin,

$$j_{Ao} = k_A(C_{Ao} - C_{Aoi}) \tag{4.77d}$$

for the amine and

$$j_{AHPo} = k_{AHP}(C_{AHPoi} - C_{AHPo}) \tag{4.77e}$$

for the penicillin-amine complex.

For steady state $j_{Pa} = j_{Ao} = j_{AHPo}$, the interfacial concentrations can be eliminated.

With the combination of these equations for the extraction rate of the penicillin acid, we obtain [322]:

$$
\begin{aligned}
-\frac{dC_P}{dt} = k_P a_p \Bigg[& C_P + 0.5\left(\frac{k_A}{k_{AHP}K_G C_H} + \frac{k_A C_A}{k_P} - C_P\right) \\
& - \left\{ 0.25\left(\frac{k_A}{k_{AHP}K_G C_H} + \frac{k_A C_A}{k_P} - C_P\right)^2 \right. \\
& \left. + \frac{k_A C_P}{k_{AHP}K_G C_H} + \frac{k_A C_{AHP}}{k_P K_G C_H}\right\}^{0.5} \Bigg]
\end{aligned}
$$

(4.79)

where a_p is the specific liquid/liquid interfacial area, k_A, k_P and k_{AHP} are the mass transfer coefficients of the amine (in the organic phase), penicillin anion (in the aqueous phase), and the amine complex (in the organic phase),

$$
K_G = \frac{C_{AHPi}}{C_H C_{Pi} C_{Ai}} \quad \text{the equilibrium constant,}
$$

C_A, C_{AHP}, C_P and C_H are the concentrations of the amine and amine complex in the organic phase and the penicillin acid anion and proton in the aqueous phase, if the reaction is controlled by the mass transfer of the components to and from the interface, where instantaneous reaction takes place, and the concentrations of the free penicillin acid in both of the phases can be neglected. In the pH range 5 to 8, in which the extraction and reextraction are performed, these assumptions are fulfilled.

The relationship (4.79) was numerically integrated to calculate the extraction of the penicillin acid anion from the organic phase. The same model was used for

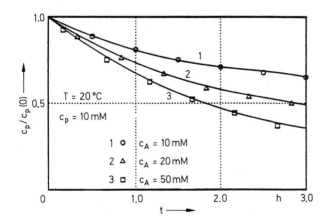

Fig. 4.43. Dimensionless penicillin concentration in the aqueous phase as a function of the extraction time at pH 6 for various amine concentrations showing the extraction kinetics [322] (——, calculated; $\bigcirc$, $\triangle$, $\square$, measured of penicillin G in an Amberlite LA-2-n-butylacetate-system ($T = 20$ °C; $C_P(0) = 10$ mM; *curve 1*, $\bigcirc$, $C_A(0) = 10$ mM; *curve 2*, $\triangle$, $C_A(0) = 20$ mM; *curve 3*, $\square$, $C_A(0) = 50$ mM)

the description of the reextraction of the penicillin acid anion from the organic phase by shifting the pH value of the aqueous phase to higher values (7.5–8).

To evaluate the kinetics of the extraction, investigations were carried out in a stirred cell at 20 °C by measuring the pencillin G concentration on-line with a polarimeter [322].

Figure 4.43 shows examples for extraction of penicillin G with LA-2 in *n*-butyl acetate at pH 6 with different amine excess and Fig. 4.44 at pH 5 with different penicillin excess. In Fig. 4.45, curves for the reextraction at pH 7.5 and 8 are shown.

For extraction, the actual concentration of penicillin acid anion C_P with regard to its initial concentration $C_P(0)$, and for reextraction with regard to the initial concentration of complex $C_{AHP}(0)$ are plotted as a function of the extraction time. The measurements were used to evaluate the parameters of the two-film model.

Equilibrium constant K_G of of the complex formation was determined as $1.25 \times 10^8 \, 1^2 \, mol^{-2}$. The specific interfacial area (0.293 m^{-1}) was given by the geometry of the stirred cell. The mass transfer coefficients were identified as $k_P = 4.5 \times 10^{-4} \, cm \, s^{-1}$, $k_A = 1.0 \times 10^{-3} \, cm \, s^{-1}$ and $k_{AHP} = 6.5 \times 10^{-4} \, cm \, s^{-1}$, where the ratio of k_A/k_{AHP} was assumed to be constant (3:2). This ratio was evaluated from the diffusion coefficients calculated according to Wilke and Chang [346].,

In all investigated extraction and reextraction processes, the measured (symbols) and with these parameters calculated (curves) data agree well, as long as the molar excess of-penicillin was less than 4 (Fig. 4.44).

From the investigations, the following conclusions can be drawn: At pH 5 the amine extractant concentration has only a slight effect on the extraction rate

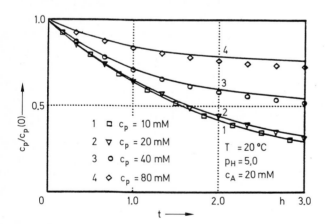

Fig. 4.44. Dimensionless penicillin concentration in the aqueous phase as a function of the extraction time at pH 5 for various penicillin concentrations showing the extraction kinetics (----, calculated; □, ▽, ○, ◇, measured) of penicillin G in an Amberlite LA-2-*n*-butyl acetate system (T = 20 °C; $C_A(0)$ = 20 mM [322]: *curve 1,* □, $C_P(0)$ = 10 mM; *curve 2,* ▽, $C_P(0)$ = 20 mM; *curve 3,* ○, $C_P(0)$ = 40 mM; *curve 4,* ◇, $C_P(0)$ = 80 mM

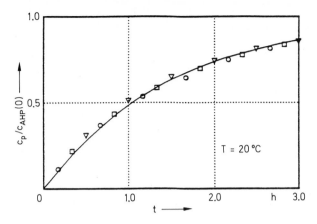

Fig. 4.45. Penicillin concentration in the aqueous phase with respect to the initial complex concentration in the organic phase as a function of the extraction time showing the kinetics of the reextraction [322]: (——, calculated; ∇, $\bigcirc$, $\square$, measured) of penicillin G from Amberlite LA-2-n-butyl acetate solution (T = 20 °C); ∇, pH 7.5, $C_A^*(0) = 10$ mM, $C_{AHP}(0) = 8.26$ mM; $\bigcirc$, pH 8.0, $C_A^*(0) = 4.6$. mM, $C_{AHP}(0) = 3.4$. mM; $\square$, pH 8.0, $C_A^*(0) = 17.6$ mM, $C_{AHP}(0) = 14.6$ mM. ($C_A^*(0)$ is the initial concentration of amine at the extraction)

(not shown), since the rate of extraction is controlled by the transport of penicillin due to the interface. At pH 6, this effect is significant because of the shift in the distribution equilibrium with increasing amine concentration at that pH value and the reduction of the penicillin concentration at the interface. With increasing amine-to-penicillin mol ratio, the extraction is enhanced (Fig. 4.43). An increase in the penicillin concentration does not influence the extraction rate as long as an excess of amine is present. With penicillin in excess, the extraction rate is reduced as a result of a reduction in the mass transfer driving force (Fig. 4.44). Also at a mole ratio of above 4, the agreement between calculated and measured concentration courses is less satisfactory.

The highest extraction rates were achieved by the amine in n-butyl acetate and isobutyl acetate, and the lowest rates with chloroform.

Extraction in a Laboratory-Scale Column and its Mathematical Simulation [323, 324, 337]. After the equilibrium and the kinetics of the extraction of penicillin were modelled in the absence and presence of amine extractants and the models were experimentally verified, the next stage towards the realization of the process was to perform the extraction in a laboratory-scale Karr column (Fig. 4.46, Table 4.32).

In order to measure the longitudinal penicillin concentration profiles in the aqueous phase, along the column, four ports – E_1 to E_4 – were used for sampling. Ports E_5 and/or E_6 (Fig. 4.46) served as tracer inlets and/or sampling tubes for residence time distribution (RTD) measurements. The continuous aqueous phase was fed into the top of the column, and the dispersed organic phase into the bottom.

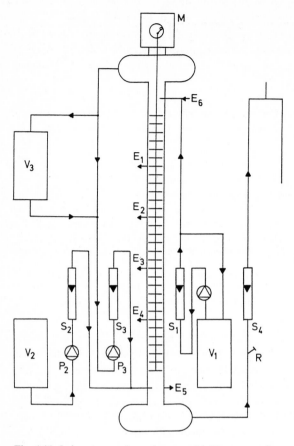

Fig. 4.46. Laboratory-scale equipment with Karr extraction column [323]: V_1, storage tank for aqueous phase; V_2, V_3, storage tanks for organic phase; P_1–P_3, pumps; S_1–S_4, flow meters; M, engine; R, valve controlling interface between the two phases; E_1–E_6, sampling and tracer injection ports

Table 4.32. Dimensions of Karr column [323]

Overall height (m)	3.75
Overall volume (l)	15.0
Active height (m)	2.2
Inner column diameter (m)	0.05
Plate number	73
Plate spacing (m)	0.0258
Diameter of plates (m)	0.0457
Number of holes per plate	16
Hole diameter (mm)	8
Free surface area of plate (%)	50
Stroke height (m)	0.018
Stroke frequency (min^{-1})	0–436

The continuous and steady-state extraction of Penicillin G by LA-2 in
n-butyl acetate was performed in the following operation range: 10 and 50 mM
LA-2; 2.5 and 5 mM Penicillin; pH 5–6.15, oscillation frequency 40–120 min^{-1},
throughput of the aqueous and organic phases: $V_a = 10–50 \, 1 \, h^{-1}$,
$V_o = 10–30 \, 1 \, h^{-1}$.

In order to be able to model the extraction process, the two-phase system in
the laboratory-scale extractor was characterized by measurements of the RTD
of both phases, the holdup of dispersed phase ε, and the droplet diameter
distributions. From the RTD, the mean residence time t of the phases, the
equivalent number of stages in a cascade model N, the Sauter droplet diameter
d_{32}, and the specific interfacial area $a_p = 6\varepsilon/d_{32}$, were evaluated for different
operating conditions [323].

The equivalent number of stages of cascade model N of the continuous phase
decreases only slightly with increasing throughput of the organic phase and the
stroke frequency f. The holdup of the organic phase ε increases considerably
with f and V_o, but only slightly with V_a. The mean residence time of the
dispersed phase t strongly increases with f, but only slightly with V_a and V_o.
With increasing f, the Sauter droplet diameter diminishes, e.g., from 1.51 mm
(f = 60 min^{-1}) to 0.86 mm (at f = 100 min^{-1}). The extraction is enhanced with
increasing f due to the decrease of d_{32}, i.e., increase of the specific interfacial area
a_p. At $\varepsilon = 0.3$ and/or f = 140 min^{-1}, flooding occurs [323].

At a high f, due to the countercurrent operation, the penicillin concentration
at the exit exceeds the equilibrium E value considerably. At f = 120 min^{-1}, 97%
and 99% of penicillin were extracted (Fig. 4.47).

In Figs. 4.48, 4.49, longitudinal profiles of the dimensionless penicillin concentration with regard to its feed concentration are shown at different f and
V_a values. The extraction is enhanced by increasing f (due to increasing a_p),

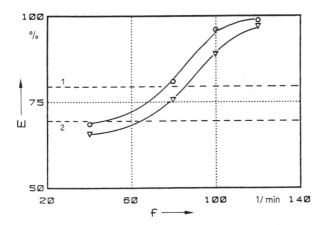

Fig. 4.47. Degrees of extraction (equilibrium) (– – –) *1*, calculated for $C_P(0) = 2.5$ mM; *2*, calculated
for $C_P(0) = 5$ mM; (——, measured at the exit of the countercurrent extractor) as a function of the
stroke frequency f at pH 5, $C_A(0) = 10$ mM [323]: $\bigcirc$, $C_P(0) = 2.5$ mM; ∇, $C_P(0) = 5$ mM

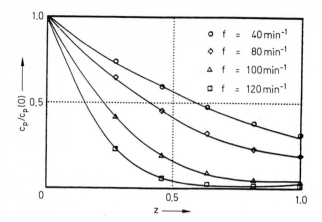

Fig. 4.48. Longitudinal profiles of penicillin concentration in the aqueous phase with regard to the feed concentration, $C_P/C_P(0)$ for various stroke frequencies [323]: $\bigcirc$, $f = 40\ \mathrm{min}^{-1}$; $\diamond$, $f = 80\ \mathrm{min}^{-1}$; $\triangle$, $f = 100\ \mathrm{min}^{-1}$; $\square$, $f = 120\ \mathrm{min}^{-1}$

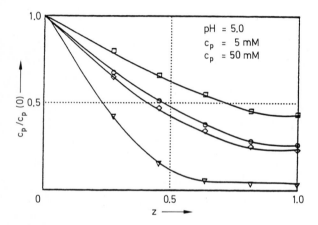

Fig. 4.49. Longitudinal profiles of penicillin concentration in the aqueous phase with regard to the feed concentration, $C_P/C_P(0)$ for various throughputs of the continuous phase and stroke frequencies (pH 5.0, $C_P = 5\ \mathrm{mM}$, $C_A(0) = 50\ \mathrm{mM}$) [323]: $\square$, $\dot{V}_a = 46\ \mathrm{l\,h}^{-1}$, $f = 60\ \mathrm{min}^{-1}$; $\bigcirc$, $\dot{V}_a = 30\ \mathrm{l\,h}^{-1}$, $f = 60\ \mathrm{min}^{-1}$; $\diamond$, $V_a = 46\ \mathrm{l\,h}^{-1}$, $f = 80\ \mathrm{min}^{-1}$; ∇, $\dot{V}_a = 46\ \mathrm{l\,h}^{-1}$, $f = 100\ \mathrm{min}^{-1}$

decreasing pH (change of equilibrium constant) and decreasing V_a (increasing amine-to-penicillin ratio).

In Fig.. 4.50, longitudinal profiles of dimensionless penicillin concentrations are shown for the reextraction of penicillin at pH 7.5, different stroke frequencies f, and initial complex concentrations $C_{AHP}(0)$. The f and $C_{AHP}(0)$ values have less effect on these profiles than on the concentration profiles for the extraction. Low f values were used, and only 80% of penicillin was attained (more or reextraction – see later).

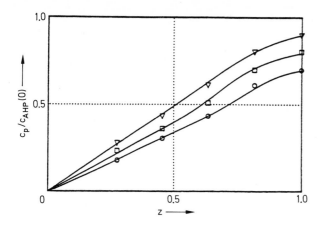

Fig. 4.50. Longitudinal profiles of penicillin concentration in the aqueous phase with regard to complex concentration in the feed, $C_P/C_{AHP}(0)$ for various initial complex concentrations in the organic solvent phase and stroke frequencies during the reextraction [323]: ∇, $C_{AHP}(0) = 3.65$ mM, $f = 80$ min^{-1}; $\square$, $C_{AHP}(0) = 8.8$ mM, $f = 80$ min^{-1}; $\bigcirc$, $C_{AHP}(0) = 8.2$ mM, $f = 60$ mM

By combining the cascade model (Sect. 2.1.5) with the kinetic model (Eq. (4.79), the following balance equation results for a single stage of the cascade [324]

$$C_{PO} - C_P = \tau k_p a_p \left[C_P + 0.5 \left(\frac{k_A}{k_{AHP} K_G C_H} + \frac{k_A C_A}{k_P} - c_P \right) \right.$$
$$- \left\{ 0.25 \left(\frac{k_A}{k_{AHP} K_G C_H} + \frac{k_A C_A}{k_P} - c_P \right)^2 \right.$$
$$+ \left. \left. \left(\frac{k_A C_{AHP}}{k_{AHP} K_G C_H} + \frac{k_A c_{AHP}}{k_P K_G C_H} \right) \right\}^{0.5} \right]. \qquad (4.80)$$

With the mass balances of the other reactants, the working point of the continuous stirred reactor (CSTR) can be calculated:

$$C_A = C_{AO} + (V_a/V_o)(C_P - C_{PO}) \qquad (4.81)$$

$$C_{AHP} = C_{AHPO} + (V_a/V_o)(C_{PO} - C_P) \qquad (4.82)$$

It is assumed that no proton gradients exist in the aqueous film at the interface.

For the evaluation of the axial concentration profiles in the countercurrent column, the reactant concentrations were calculated in the cascade of the CSTRs by assuming the outlet concentrations of amine and amine complex in the solvent from the first stage, and from these the mass transfer coefficients were identified by minimizing the differences between calculated and measured concentrations [337]. The same model was used for describing the reextraction of penicillin by considering the different fluxes of the components.

For the simulation of the extraction, measured equilibrium data, kinetic constants, the balance Eqs. (4.79–4.81), and the characteristics of the two-phase system (Table 4.33) which were evaluated by separate measurements, were used.

In order to take the low specific interfacial area a_p at the bottom of the column (feed of the organic phase) into consideration, the a_p value was reduced in the lowest tree cells to 75, 50 and 25% of its original value.

The mass transfer coefficients were identified by means of the measured longitudinal concentration profiles of penicillin and by assuming that ratio k_A/k_{AHP} is constant and obtained the following values:

$$k_P = 1.2 \times 10^{-3} \ cm \ s^{-1}$$

$$k_A = 2.2 \times 10^{-3} \ cm \ s^{-1}$$

$$k_{AHP} = 1.4 \times 10^{-3} \ cm \ s^{-1}$$

Figures 4.51, 4.52 represent measured (symbols) and calculated (curves and points) longitudinal concentration profiles for extraction of penicillin with LA-2 in n-butylacetate at pH 5 and its reextraction at pH 9.

In the investigated cases, the agreement between calculated and measured longitudinal concentration profiles and the final degrees of extraction are excellent. Only at high amine concentrations, are there some differences between them that are probably due to the slower dispersion rate of the organic solvent phase in the column because of its higher viscosity. The reextraction of penicillin was described by the same model and model parameters. However, because of the different hydrodynamic conditions (lower stroke frequency) in the column, the droplet size is larger.

The simulations of the extraction process indicate that at low penicillin concentrations (amine excess), the reaction kinetics dominate, and at high penicillin concentrations the equilibrium significantly influences the extraction.

To compare the performance of the laboratory-scale Karr column with other columns, investigations were carried out also with a laboratory-scale Kühni extractor (Fig. 4.53) and a laboratory-scale pulsated perforated plate column (Fig. 4.54). The most important data of the three columns are given in Table 4.34 [331]

In the three columns, the same operation conditions – mean residence time and throughput ratio of the phases (2 : 1) as well as pH (5.12), penicillin concen-

Table 4.33. Experimentally determined model parameters of the laboratory scale Karr column

$f(min^{-1})$	40	80	100	120
ε	0.0036	0.046	0.082	0.148
$d_{32}(mm)$	1.5	1.15	0.9	0.8
$a_p(cm^{-1})$	1.5	2.6	6.0	13.0
N	19	16	14	12
$\tau(s)$	730	720	695	645

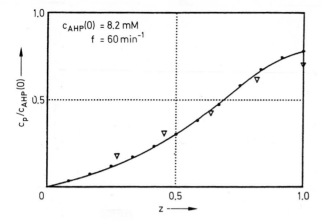

Fig. 4.51. Longitudinal profiles of the penicillin concentration in the aqueous phase with regard to the feed concentration, $C_P/C_P(0)$ at different stroke frequencies, pH = 5.0; $C_A(0)$ = 10 mM [324]: calculated (*curve*) and measured (*symbols*) data: $\square$, f = 40 min^{-1}, $\bigcirc$, f = 80 min^{-1}, $\triangle$, f = 120 min^{-1}

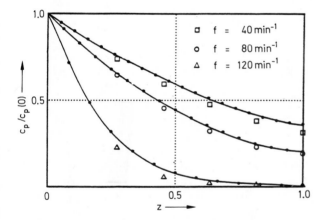

Fig. 4.52. Longitudinal profile of penicillin concentration in the aqueous phase with regard to the complex concentration in the feed, $C_P/C_{AHP}(0)$ during the reextraction of penicillin [324]: comparison of calculated (*curve*) and measured (*symbols*) data: $C_{AHP}(0)$ = 8.2 mM, f = 60 min^{-1}

tration (C_P = 5 mM), and amine concentration (C_A = 20 mM) in feed – were used.

The columns were characterized by measurements of N, d_{32}, ε, a_p and longitudinal concentration profiles of penicillin in the aqueous phase. The degrees of extraction and the longitudinal concentration profiles were calculated by means of Eqs. (4.79–4.81) and compared with the measurements [332]. Again, satisfactory agreement between calculated and measured profiles was observed.

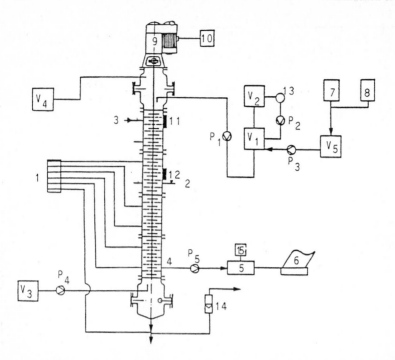

Fig. 4.53. Laboratory-scale equipment with Kühni extractor [331]: P_1-P_5, pumps; V_1, V_2, storage tanks for aqueous phases; V_3, storage tank for organic phase; V_4, storage tank for loaded organic phase; V_5, storage tank; *1*, sampling ports; *2*, cock for the determination of holdup; *3*, tracer injection port; *4*, sampling port; *5*, photometer; *6*, recorder; *7*, *8*, distillation unit (not used in these investigations); *9*, engine; *10*, rotation speed meter; *11*, *12*, windows for flash photography; *13*, heat exchanger; *14*, flow meter; *15*, printer

In Fig. 4.55, the longitudinal concentration profiles in the tree columns measured under the conditions given above, are compared. It is obvious from Fig. 4.55 that the performance of the Kühni extractor is the highest, and that the performances of PPP- and Karr columns are very similar.

Extraction in a Pilot-Plant-Scale Column and its Mathematical Simulation [325–327, 338]. After penicillin extraction in a laboratory-scale column was established, the process was transferred into a pilot plant. In Fig. 4.56, the pilot plant columns are shown, and in Table 4.35, the explanation of Fig. 4.56 and the column data are given.

First, penicillin model media were applied to evaluate the performance of the column. Again, LA-2 in *n*-butyl acetate was used as extractant. Penicillin concentrations up to $30 \, \mathrm{g \, l^{-1}}$, throughputs of the aqueous phase of $100 \, \mathrm{l \, h^{-1}}$ and aqueous-to-organic phase throughput ratios of three, and very high degrees of extraction (99%) were achieved at high stroke frequencies, slight carrier excess (1.5–2.0), and pH values between 5.0 and 5.2.

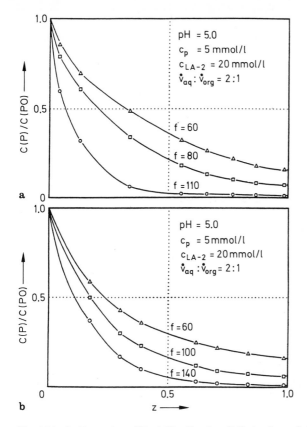

Fig. 4.54a, b. Extraction of Penicillin G: **a** in a Kühni column; **b** in a pulsated perforated column. Dimensionless penicillin concentration as a function of the dimensionless length of the column at different rpm and/or oscillation frequencies during the extraction with LA-2 in *n*-butylacetate [331].

Table 4.34. Comparison of the data three columns [331]

Data of columns	Karr	PPP	Kühni
Active length (mm)	2200	2235	1440
Active volume (ml)	4200	18500	10540
Internal diameter (mm)	50	100	100
Number of trays	73	20	24 (cells)
Hole dia. of perforated plates	8	2	
Amplitude (mm)	0–18	0–22	
Free surface area of trays (%)	50	20	30
Stroke frequency (min^{-1})	0–436	0–160	
Impeller speed (min^{-1})			0–180
$V_a/V_o(L\ h^{-1}/L\ h^{-1})$	20/10	86/43	50/25

PPP: pulsated perforated plate column

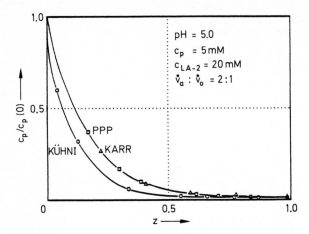

Fig. 4.55. Comparison of the dimensionless penicillin concentrations in the Kühni (○), PPP (□) and Karr (△) columns as functions of the dimensionless column length during the extraction of penicillin G with LA-2 and *n*-butylacetate [331]

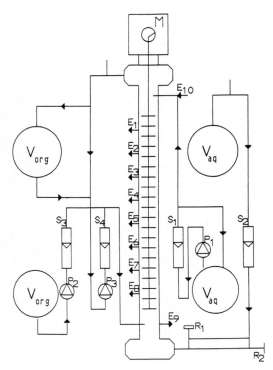

Fig. 4.56. Pilot plant Karr extracation column (see Table 4.35 [325]) E_1 to E_{10} sampling ports, P_1 to P_3 pumps, S_1 to S_4 flow meters, V_{org} and V_{aq} storage vessels for organic and aqueous phases, M motor, R_1 and R_2 control valves

The increase of the stroke frequency from $f = 60$ min^{-1} to 100 min^{-1} decreases the Sauter diameter from $d_{32} = 2.62$ mm to 1.25 mm, increases the holdup from $\varepsilon = 0.04$ to 0.13, and the specific interfacial area by a factor of five. At a low stroke frequency, the longitudinal holdup distribution in the columns

Table 4.35. Explanation of Fig. 4.56 and column data

Explanation of Fig. 4.56	
E_1–E_{10}	sampling ports
M	drive mechanism
P_1–P_3	pumps
R_1, R_2	control valves
V	liquid reservoir

Data of Karr-column	
overall height	7.6 m
overall volume	46 l
active exchange height	4 m
active volume	22 l
internal diameter	0.08 m
numbers of sieve plates	108
distance between sieve plates	0.025 m
amplitude	0.016 m
stroke frequency	0–325 min^{-1}
drive power	0.6 kW
max. throughput	220 l h^{-1}

are uniform. At high f values, the high holdup is reduced considerably at the bottom (feed of the organic phase) of the column [325].

High degrees of penicillin extraction (98–99%) and penicillin enrichments of three (at $V_a/V_o = 3$) were attained also by extraction with LA-2 in n-butyl acetate at pH 5 from fermentation media [326].

Comparison between the holdups and the longitudinal concentration profiles of penicillin from model media and from fermentation broths indicate differences (Fig. 4.57). For higher degree of extraction in model media, however, the interfacial area cannot be held responsible, since it is higher in the fermentation broths than in the model media.

The difference is mainly caused by the coextraction of organic acids from the fermentation media.

The enhancement of the extraction by increasing the stroke frequency is markedly limited by the low flooding point, which is lower than in model media.

The reextraction was carried out with phosphate buffer at pH values above 7.5. The reextraction system has a higher tendency for emulsion formation than the extraction system. Therefore, the critical stroke frequency is lower than that for extraction. The pH value drifts along the column. The low frequency and the pH shift are responsible for the relatively low degrees of reextraction (80-86%). With a pH control along the column, the degree of extraction can be increased to 88%. With a V_o/V_a-ratio of three, a penicillin enrichment of three can be attained. Penicillin losses are below 1%. No amine losses could be detected [326] after seven weeks of operation.

The longitudinal concentration profiles were calculated for different operating conditions and compared with the measurements [327] by means of the

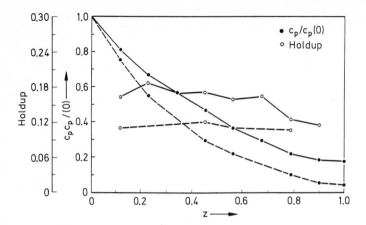

Fig. 4.57. Comparison of longitudinal profiles of penicillin concentration (●) and holdup (○) in the pilot plant Karr column with model medium (– – – –) and fermentation broth (——) [326]: $V_a = 100\,\mathrm{l\,h^{-1}}$; $V_o = 50\,\mathrm{l\,h^{-1}}$; $C_P(0) = 5\,\mathrm{g\,l^{-1}}$; $C_A(0) = 20\,\mathrm{g\,l^{-1}}$; $f = 60\,\mathrm{min^{-1}}$

balance equations (Eqs. 4.79–4.81) and the experimental evaluated data of the two-phase system in the column (Table 4.36).

The agreement between the measured and calculated profiles is satisfactory (Figs. 4.58, 4.59). The identified model parameters are: $k_P = 1.2 \times 10^{-3}\,\mathrm{cm\,s^{-1}}$, $k_A = 2.2 \times 10^{-3}$, $k_{AHP} = 1.4 \times 10^{-3}\,\mathrm{cm\,s^{-1}}$. These model parameters are identical to those that were obtained in the bench column.

In Table 4.37, the specific interfacial areas, ascertained in the laboratory-scale and pilot-plant-scale-columns, are compared.

In spite of the higher specific interfacial area of the fermentation media in the pilot plant column, the degrees of extraction are somewhat less than in the model media due to the coextraction which was not taken into account in the calculations.

These calculations indicate that the simulation of the extraction process can be performed by the model equations (Eqs. 4.79–4.81) for laboratory-scale and pilot-plant-scale columns. For the layout of the columns only the hydrodynamical parameters are needed, which, unfortunately, cannot be calculated yet on a theoretical basis.

Table 4.36. Data used for the calculations of the longitudinal concentration profiles during the extraction of penicillin from the model media in the pilot plant column [327]

V_a $\mathrm{l\,h^{-1}}$	V_o $\mathrm{l\,h^{-1}}$	f $\mathrm{min^{-1}}$	d_{32} mm	ε 1	a_p $\mathrm{cm^{-1}}$	τ s	N 1
50	25	60	2.1	0.037	1.10	1525	10
50	25	80	1.59	0.057	2.28	1494	12
50	25	100	1.25	0.077	4.02	1462	14
100	25	100	1.13	0.131	8.0	688	23
100	50	40	2.22	0.156	5.0	670	15
100	50	60	1.73	0.160	6.60	670	16

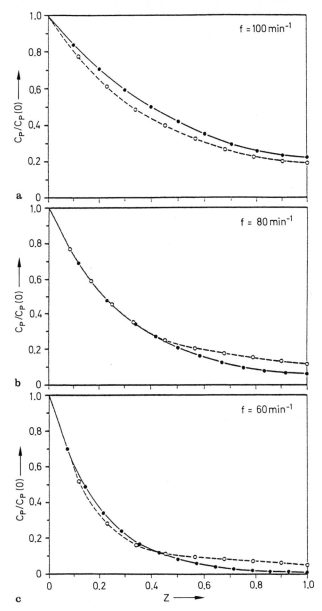

Fig. 4.58. Comparison of the measured and calculated longitudinal penicillin concentration profiles in the pilot plant Karr column at different stroke frequencies with model media [327]: f = 60 min⁻¹; **b** f = 80 min⁻¹; **c** f = 100 min⁻¹; – – – measured, — calculated

Problems of Reextraction [328, 339]. Until now, extraction and reextraction were considered as two separate processes. They are, however, interrelated. Therefore, these two processes are investigated together. At first, reextraction is considered and then extraction is coupled with reextraction.

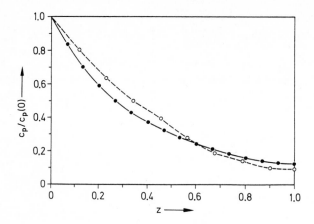

Fig. 4.59. Comparison of measured and calculated longitudinal penicillin concentration profiles in the pilot plant Karr column with fermentation broth [327]: – – – – measured, — calculated

Table 4.37. Specific interfacial area a_p (cm^{-1}) measured during the extraction in laboratory-scale and pilot-plant-scale columns [327]

f (min^{-1})	40	60	80	100	120
laboratory column	1.5	—	2.6	6.0	13.0
pilot plant (model)	—	1.1	2.28	4.02	—
pilot plant ferment. medium	5.0	6.6	—	8.0	—

Extraction is characterized by the degree of extraction E, reextraction by the degree of reextraction E_R, and the extraction-reextraction process by the overall degree of extraction E_G. (The organic solvent phase that is in equilibrium with the aqueous broth is separated from the broth and equilibrated with the reextraction buffer. The degree of extraction-reextraction, E_G, is the amount of the penicillin in the buffer phase with regard to its amount in the broth.) Several amine extractants were applied. The investigations were usually carried out under the following conditions: the volumes of the organic phase (V_o) and aqueous phase (V_a) were equal. Extractions were performed at pH 5, with $C_P = 10$ mM and $C_A = 0, 10, 20, 40,$ and 80 mM, and re-extraction at pH 5.0, 5.5, and 8.5–9.0.

The investigations of the extraction of penicillin with amine extractants indicated that the degree of extraction improves with decreasing pH value and increasing amine concentration. In Fig. 4.60, the degree of reextraction is shown as a function of the pH value at different concentrations of amine extractants (LA-2). As expected, the degree of re-extraction improved with increasing pH value and decreasing amine concentration. The degrees of re-extraction are

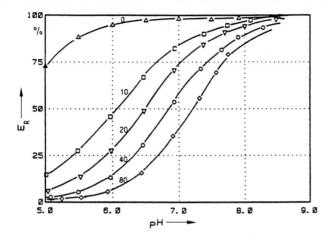

Fig. 4.60. Degree E_R of reextraction as a function of pH value at different LA-2 (mM) concentrations; $C_P = 10$ mM [328]

adequate at pH 8 for amine excesses up to two. Above that, they fell with increasing amine concentration. Since the amine concentration has opposite effects on extraction and reextraction, one would expect a maximum of E_G at intermediate amine concentrations. However, no maximum of E_G is obtained. With increasing amine concentration, E_G improves and approaches a constant value (Fig. 4.61). At LA-2 excess ($C_A/C_P = 4$) and pH 8, $E_G = 90\%$ can be obtained. However, a high pH is detrimental for the stability of penicillin and for phase separation (Fig. 4.62). Therefore, the recommended conditions are pH

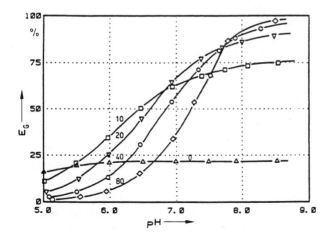

Fig. 4.61. Degree E_G of extraction and reextraction as a function of the pH value at different LA-2 (mM) concentrations; $C_P = 10$ mM [328]

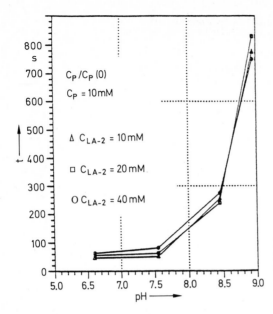

Fig. 4.62. Phase separation time as a function of pH value at different carrier concentrations [328]

5 and amine excesses of two to four for the extraction step, and pH 7.5 for re-extraction. With increasing penicillin concentration (and at constant amine concentration), the phase separation time also increases (Fig. 4.63). For example, at a penicillin concentration $C_P = 54$ mM and pH 7.75, the separation time is 10 min. The phase separation time is also influenced by the buffer used for reextraction.

Besides the secondary amine LA-2, several other commercially available amine extractants were also tested for penicillin extraction. In [339], 21 of them are listed, together with the corresponding E (pH 5.0) and E_R (pH 7.5) values. In Table 4.38, the equilibrium constants and the degrees of extraction E and E_R of the best nine of them (chemically pure components) are compiled.

The amines were dissolved in 25 ml n-butylacetate and contacted with 25 ml aqueous penicillin solution. Penicillin was then reextracted from the organic phase with 25 ml buffer solution. The secondary amine, diisotridecylamine ($C_{13}H_{27}$-NH-$C_{13}H_{27}$, MW 381) turned out to be the best amine extractant. It is soluble in n-butylacetate and insoluble in water, has a density of 0.830 g l^{-1} and a kinematic viscosity of 34 cP. Its has little tendency to form stable emulsions.

In Fig. 4.64, the degree of extraction of penicillin with diisotridecylamine (DITDA) is plotted as a function of the pH value at different DITDA concentrations (mM) with $C_P = 5$ mM. At pH 5, and an amine-to-penicillin mol ratio of two, E = 95% was obtained.

Both LA-2 and DITDA are suitable as amine extractants and yield about the same E and E_R values. The equilibrium constants of their reaction with penicillin are comparable: K_G (DITDA) $= 2.10 \times 10^8$ l^2 mol^{-2}; K_G (LA-2) $= 1.25 \times 10^8$ l^2 mol^{-2}.

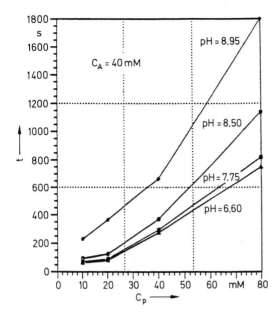

Fig. 4.63. Phase separation time as a function of penicillin concentration at different pH values [328]

Table 4.38. Amine extractants and their equilibrium constants K_G and degrees of extraction E (pH 5.0) and reextraction E_R (pH 7.5)

Extractant	Equil. const. $K_G \times 10^{-7}$ $(L^2 mol^{-2})$	E (pH 5.0) (%)	E_R (pH 7.5) (%)
Diisotridecylamine	21	83.0	94.3
Trioctyldecylamine	39.1	72.2	95.2
Dicocos amine	37.3	95.3	62.5
Dimethyl coconut oil amine	3.59	78.4	93.5
Dimethyllaurylamine	2.89	76.4	94.7
Dimethylstearylamine	3.65	79.3	92.2
Dimethylsoyamine	3.85	80.0	94.1
Dimethyldodecylamine	3.21	76.0	93.2
Dimethyltetradecyl amine	3.27	77.9	92.2

A comparison of the extraction courses of penicillin G with LA-2 and DITDA in *n*-butylacetate in a stirred cell is shown in Fig. 4.65. The extraction rate is somewhat higher with DITDA than with LA-2.

Since the economy of the recovery is also influenced by the cost of the carrier, their prices should be compared: Amberlite LA-2 (Roehm and Haas) costs 72 cents per l, while Hoe F 2562 DITDA (Hoechst) costs 3.3 cents per l. Due to a 20-fold lower cost and its high purity (it is free of solvents), the DITDA is an ideal amine extractant for penicillin.

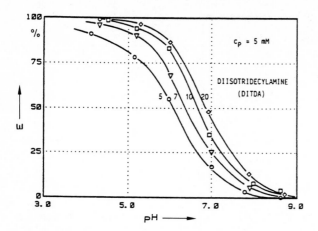

Fig. 4.64. Degree E of extraction of penicillin G as a function of the pH value with diiso-tridecylamine (DITDA)-*n*-butylacetate system at different DITDA concentrations (mM) [328]. $C_P = 5$ mM, pH 5

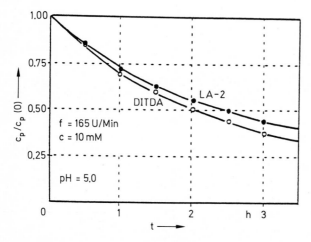

Fig. 4.65. Dimensionless penicillin concentration with regard to its initial value $C_P/C_P(0)$ as a function of the extraction time using 10 mM LA-2 and/or DITDA in *n*-butyl acetate at pH 5.0 and impeller speed 165 rpm in stirred cell [328]. $C_P(0) = 10$ mM; $V_a = V_o = 160$ ml; buffer: Na citrate

Kinetics of Reextraction [329, 339]. Since the extraction of penicillin with amine extractant is an equilibrium reaction Eq. (4.69), Eqs. (4.70–4.75) are applicable. Again, the two-film model was used (Fig. 4.66). The following balances hold:

$$j_P = k_P (C_{Pi} - C_P) \tag{4.82a}$$

$$j_A = k_A (C_{Ai} - C_A) \tag{4.82b}$$

$$j_{AHP} = k_{AHP} (C_{AHP} - C_{AHPi}). \tag{4.82c}$$

$$AHP_o \rightleftharpoons A_o \ + \ P_a^- \ + \ H_a^+$$

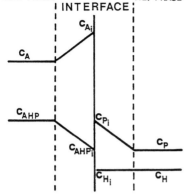

Fig. 4.66. Schematic concentration profiles of the components at the interface according to the two-film model during the reextraction of penicillin [329]

For steady-state conditions ($j_A = j_P = j_{AHP}$), one obtains:

$$\frac{dc_P}{dt} = a_p k_P \left[\left\{ \frac{B}{2} + \left(\frac{B^2}{4} - C \right)^{0.5} \right\} - C_P \right] \tag{4.83}$$

with

$$B = \frac{C_H C_A k_A k_{AHP} + k_P k_A K_G - C_H C_P k_P k_{AHP}}{C_H k_P k_{AHP}} \tag{4.83a}$$

and

$$C = \frac{k_{AHP} C_{AHP} K_G k_A + k_P C_P K_G k_A}{k_{AHP} k_P C_H}, \tag{4.83b}$$

if the reaction is controlled by the mass transfer of the components to and from the interface, where the instantaneous reaction takes place, and the concentrations of the free penicillic acid in both of the phases can be neglected.

The solution of Eq. (4.83) is obtained by numerical integration.

The re-extraction from the organic phase was carried out in a stirred cell with phosphate buffer at $T = 20\ °C$.

In Fig. 4.67, the dimensionless penicillin concentration is plotted as a function of time at pH 7.0 with the initial amine penicillin complex concentration C_{AHP} (0) as the parameter. C_{AHP} (i.e., the amine concentration) has considerable influence on the extraction rate. With increasing amine concentration the re-extraction rate diminishes. However, this effect diminishes with increasing pH value, and at pH 9 it disappears [329]. On the other hand, the pH effect is significant at C_{AHP} (0) = 2.8 mM (Fig. 4.68), but it diminishes with increasing amine concentration (Fig. 4.68). In Fig. 4.67, 4.68, both the measured (symbols) and calculated (curve) data are plotted and compared. The agreement is adequate. These measurements were used to identify the mass transfer coefficients.

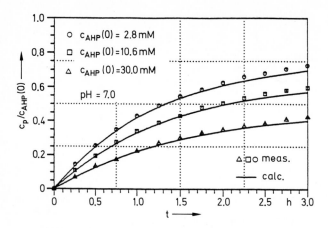

Fig. 4.67. Dimensionless penicillin concentration with regard to the initial complex concentration $C_P/C_{AHP}(0)$ as a function of the extraction time during the reextraction from LA-2-n-butylacetate solution at different initial complex concentrations and at pH 7.0: ($\bigcirc$) $C_{AHP}(0) = 2.8$ mM; ($\square$) $C_{AHP}(0) = 10.6$ mM; ($\triangle$) $C_{AHP}(0) = 30.0$ mM [329]

Again by assuming a constant k_A/k_{AHP} ratio:

$$k_P = 5.2 \times 10^{-4} \text{ cm s}^{-1},$$

$$k_A = 1.1 \times 10^{-4} \text{ cm s}^{-1}, \quad \text{and}$$

$$k_{AHP} = 7.4 \times 10^{-4} \text{ cm s}^{-1} \quad \text{were obtained.}$$

The equilibrium constant was $K_G = 3.0 \times 10^{-9} \text{ mM}^2$.

The measurements indicate that the extraction is enhanced with increasing pH and attains a constant value at pH 9. At this pH, neither pH nor the initial penicillin-amine complex concentration influences the extraction rate. This behavior can be described well with the mathematical model developed in [329].

Reextraction in a Laboratory-Scale Column and its Mathematical Simulation [330–332, 339]. The reextraction of penicillin G and V from the penicillin-LA-2 complex in butyl acetate by phosphate buffer was investigated in a laboratory-scale Karr column (Fig. 4.46) in countercurrent mode.

Initially, penicillin was extracted from a model medium or cell-free fermentation medium with LA-2 in butyl acetate at pH 5.0. This penicillin-LA-2 complex was reextracted from the organic phase with a buffer solution.

The residual penicillin was then completely removed from the organic phase by means of the buffer solution at pH 8.5 and the organic phase was reused again. The holdup, the droplet-size distributions, and the longitudinal penicillin concentration in the aqueous phase were measured.

With phosphate buffer, the phase separation was adequate below pH 8. At pH 9, no steady-state operation was possible.

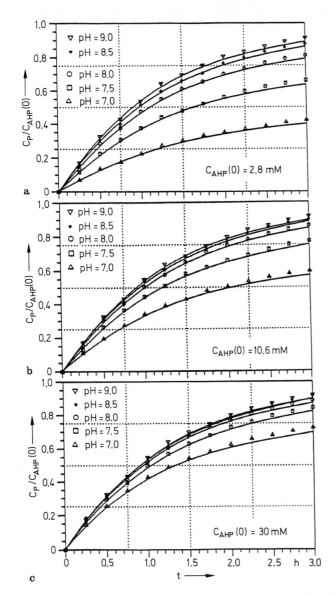

Fig. 4.68a–c. Dimensionless penicillin concentration with regard to the initial complex concentration $C_P/C_{AHP}(0)$ in the organic phase as a function of the extraction time during the reextraction from LA-2-n-butylacetate solution at different pH values: **a** $C_{AHP}(0) = 2.8$ mM; **b** $C_{AHP}(0) =$ 10.6 mM; **c** $C_{AHP}(0) = 30.0$ mM [329]

In Fig. 4.69, longitudinal pH and concentration profiles in the aqueous phase are shown at $C_{AHP}(0) = 9$ mM, $V_o = 10\,l\,h^{-1}$, $V_a = 20\,l\,h^{-1}$, and at different stroke frequencies f. The pH value diminishes only slightly because of the relatively high throughput ratio ($V_a/V_o = 2$). With a decreasing throughput

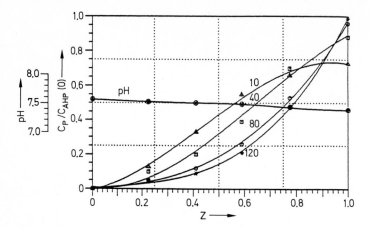

Fig. 4.69. Longitudinal profiles of the penicillin concentration in the aqueous phase with regard to the complex concentration in the feed of the organic phase $C_P/C_{AHP}(0)$ and the pH value during the reextraction of penicillin at four different stroke frequencies [330]: $C_{AHP}(0) = 9$ mM; $V_o = 10\,l\,h^{-1}$; $V_a = 20\,l\,h^{-1}$; $\bigcirc$, pH; $\triangle$, f = 10 min^{-1}; $\square$, f = 40 min^{-1}; $\bigcirc$, f = 80 min^{-1}; $*$, f = 120 min^{-1}

ratio, the pH value falls along the column more and more. This shifts the extraction equilibrium to the formation of the complex, and reduces the degree of reextraction. Increasing stroke frequency enhances the extraction in the lower part of the column (z > 0.5). At high pH values, the longitudinal pH profiles are mostly a result of the coextraction of OH$^-$ ions. This shifts the equilibrium and reduces the reextraction rate. Therefore, the pH value has no significant influence on the longitudinal concentration profiles of penicillin (Fig. 4.70).

Penicillin V was extracted from the fermentation broth after the removal of mycelium by a centrifugal separator, setting the pH value at 5.0, and removing the precipitated proteins. At pH 5.2, 99% of the penicillin was extracted. The reextraction was carried out at pH 7.65 with $C_{AHP} = 9.2$ mM, $V_a = 10\,l\,h^{-1}$, $V_o = 20\,l\,h^{-1}$, and f = 100 min^{-1}; 97% of the penicillin was reextracted. No phase separation problems occurred.

The main problem in extraction is caused by precipitated protein, which limits the stroke frequency to fairly low values. This problem can be avoided if the fermentation is stopped at about 190 h, before the protein content in the broth increases by a factor of three to four up to 260 h. The reextraction of penicillin V from such a fermentation broth is shown in Fig. 4.71. The longitudinal penicillin V concentrations in the aqueous phase are shown at two different times (t = 190 and 260 h). Due to the lower protein content of the broth that was harvested after a shorter fermentation time,t he extraction can be carried out at higher stroke frequencies (f = 85 min^{-1}) than the one harvested after a longer fermentation time and having higher protein content (f = 70 min^{-1}). At f = 85 min^{-1}, 98% of the penicillin was extracted. This is considerably higher than that (82%) extracted at f = 70 min^{-1}.

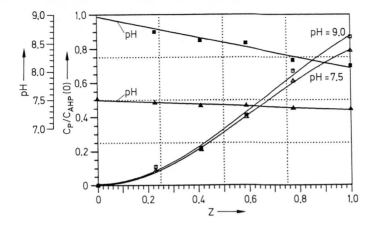

Fig. 4.70. Longitudinal profiles of the penicillin concentration with regard to the complex concentration in the feed of the organic phase $C_P/C_{AHP}(0)$ during the reextraction of penicillin at two different pH values: △ pH = 7.5 and □ 9.0 [330]: $C_{AHP}(0) = 9$ mM, $V_o = V_a = 15 \, l h^{-1}$, $f = 100 \, min^{-1}$ ▲ and ■ pH values

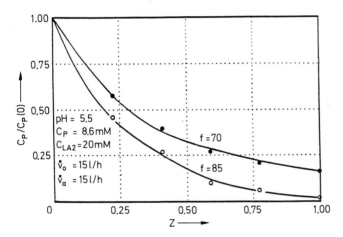

Fig. 4.71. Longitudinal profiles of the penicillin concentration with regard to the complex concentration in the feed of the organic phase $C_P/C_{AHP}(0)$ during the reextraction from two different broths [330]: ○, broth at fermentation time of 190 h; ●, broth at the fermentation time of 260 h

All these laboratory-scale reextraction processes were carried out in a Karr column. In order to investigate the influence of the column construction on the reextraction performance, the reextraction of penicillin G was also investigated in two other columns: a Kühni extractor (Fig. 4.53) and a pulsated perforated plate (PPP) column (Fig. 4.15). The characteristics of the three (Karr, Kühni, and PPP) columns are given in Table 4.34 [331].

In Figs. 4.72, 4.73, the longitudinal concentration profiles of penicillin in the aqueous phase are shown for reextraction at pH 7.5 in PPP and Kühni columns at different pulsation frequencies and/or impeller speeds. The reextraction is enhanced with increasing frequency and/or impeller speed. The maximal performances of the three columns, which are near their flooding points, are compared in Fig. 4.74. The Kühni column exhibits the best performance.

In order to find out the cause for the variations in the performances of these columns, their holdups and Sauter droplet diameters were determined. The

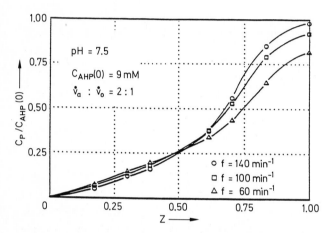

Fig. 4.72. Reextraction of penicillin G in a pulsated perforated plate column at different pulsation frequencies. Dimensionless concentration with regard to the complex concentration in the feed of the organic phase as a function of the dimensionless column length [331]

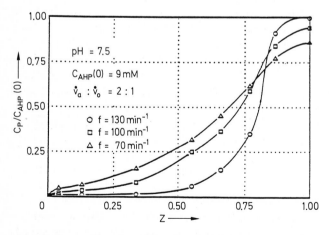

Fig. 4.73. Reextraction of penicillin G in a Kühni column at different impeller speeds. Dimensionless penicillin concentration with regard to the complex concentration in the feed of the organic phase as a function of the dimensionless column length [331]

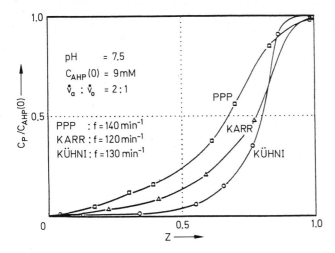

Fig. 4.74. Reextraction of penicillin G in different columns. Comparison of the profiles of the dimensionless penicillin concentration with regard to the complex concentration in feed of the organic phase in the Kühni (○), Karr (△) and PPP (□) columns as a function of the dimensionless column length near to the flooding point [331]

lowest Sauter diameters were found in the Kühni column. With increasing impeller speed, the holdup in the Kühni column increased more and attained higher values near the flooding point, than in the other columns. This holds true for extraction as well as for reextraction. Lower Sauter droplet diameters and higher holdups yield higher specific interfacial areas as well as higher extraction and reextraction rates in the Kühni extractor.

The applicability of a mathematical model developed for reextraction in the PPP and Kühni extractors was also investigated. In this model, the systems were characterized by measurements of the mean residence time (τ), N, d_{32}, ε, a_p, and the longitudinal concentration profiles of penicillin in the aqueous phase. The degrees of extraction and the longitudinal concentration profiles were calculated by the combination of a cascade model (Chapter 2.1.5) with the kinetic equation (Eq. 4.83) [332]. The mass transfer coefficients were evaluated by means of measurements (Table 4.39).

The simulations of the reextraction in the PPP column at different pulsation frequencies (Fig. 4.75) and in the Kühni extractor at different impeller speeds (Fig. 4.76) by means of this model yielded adequate results.

One can state that the model is able to describe the extraction and reextraction of penicillin in all three columns.

Extraction in a Mixer-Settler [333, 339]. The mixer-settler is the most popular extractor system [347]. Therefore, investigations were performed with a four-stage laboratory mixer-settler (Type DN 80 of QVF, Wiesbaden) with an overall phase-throughput rate of 60 to 80 l h^{-1}. Penicillin was produced in a 250 l

Table 4.39. Operating data and evaluated model parameters for the simulation of the reextraction of penicillin G in the three laboratory-scale columns [330]

V_a $l\,h^{-1}$	V_o $l\,h^{-1}$	pH	f min^{-1}	k_P, k_A, k_{AHP} $cm\,s^{-1} \times 10^{-3}$	N	d_{32} mm	ε	τ s	a_p cm^{-1}
Karr									
15	15	7.5	110	1.3, 1.6, 1.1	10	1.16	0.078	929	4.4
15	15	7.5	90	1.3, 1.6, 1.1	10	1.36	0.052	956	2.4
15	15	7.5	60	1.3, 1.6, 1.1	10	1.68	0.038	970	1.4
20	10	7.5	120	1.3, 1.6, 1.1	19	0.82	0.044	723	3.4
20	10	7.5	80	1.3, 1.6, 1.1	19	1.02	0.036	729	2.2
20	10	7.5	40	1.3, 1.6, 1.1	19	1.88	0.028	735	0.9
15	15	7.5	100	1.3, 1.6, 1.1	10	1.26	0.048	960	2.4
15	15	8.6	100	1.3, 1.6, 1.1	10	1.18	0.047	961	2.5
PPP									
86	43	7.5	140	1.8, 1.6, 1.1	6	0.92	0.07	720	4.9
86	43	7.5	100	1.8, 1.6, 1.1	6	1.21	0.056	731	3.0
Kühni									
50	25	7.5	100	1.5, 1.6, 1.1	6	0.52	0.13	660	17.2
50	25	7.5	130	1.5, 1.6, 1.1	6	0.75	0.098	686	8.7

stirred tank bioreactor. The mycelium-free broth was extracted at pH 5 with LA-2 in *n*-butyl acetate at different impeller speeds (f) in countercurrent operation (Fig. 4.77). After the fourth stage, more than 90% of the penicillin was extracted from the broth. The extraction was enhanced with increasing impeller speed. As was expected, it was possible to extract more penicillin with a throughout ratio of one (aqueous phase to organic phase) than with a ratio of two. The highest extraction rate (99% of penicillin extracted) was obtained with a throughput ratio of 1:1 at high impeller speeds (Fig. 4.77). Reextraction of penicillin G from the complex-containing organic phase was carried out in the same mixer-settler with phosphate buffer at pH 7.5 and/or carbonate buffer at 7.2. Depending on the buffer type and phase throughput ratio, 70 to 95% of the penicillin was reextracted (Fig. 4.78).

Extraction in Centrifugal Extractors [334, 339]. Centrifugal extractors are used for penicillin recovery on the commercial scale (Chapter 3.1) [307–311, 348]. Therefore, laboratory-, and pilot-plant-scale centrifugal extractors were used for the extraction of penicillin from model media and mycelium free fermentation broths at pH 5 with LA-2 in *n*-butyl acetate. The reextraction was carried out at pH 7.5 with different buffer solutions.

For laboratory investigations, the separator SA 01 of Westfalia (Oelde) was used, which is a single-stage extractor. Both phases enter and leave the extractor at the top. Preliminary tests indicated that a control disk of 55 mm in diameter is adequate for the separation.

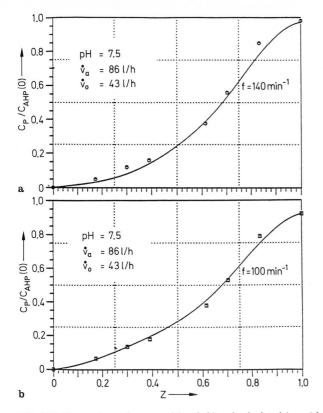

Fig. 4.75. Comparison of measured (*symbols*) and calculated (*curve*) longitudinal profiles of dimensionless penicillin concentration with regard to the complex concentration in the feed of the organic phase in pulsated perforated plate column at different pulsation frequencies [332]

For the extraction, two stages of the separator SA 01, model medium $(C_P(0) = 5\,\text{mM};\ C_A(0) = 20\,\text{mM})$, and fermentation broth of penicillin V $(C_P(0) = 5\,\text{mM};\ C_A(0) = 20\,\text{mM})$ were used. The operating conditions $(V_a = 15\,\text{l}\,\text{h}^{-1}$ and $V_o = 7.5\,\text{l}\,\text{h}^{-1})$ were kept constant.

Figure 4.79 shows the dimensionless penicillin concentrations with regard to their feed concentration after the first and second stages, as well as the pH value in the aqueous phase as a function of the extraction time. After the first stage, 87.3% and 91.8% of the penicillin was extracted from the model medium and from the broth, respectively. Fresh solvent was used in the second stage. After the second stage, the extraction from both of the solutions was practically complete. The reextraction from the complex-containing organic phase $(V_o = 15\,\text{l}\,\text{h}^{-1})$ was performed with fresh phosphate buffer solutions $(V_a = 7.5\,\text{l}\,\text{h}^{-1})$ in both of the stages (Fig. 4.80). With $C_P(0) = C_A(0) = 100\,\text{mM}$ and $V_a = V_o = 15\,\text{l}\,\text{h}^{-1}$, 98% of the penicillin was extracted and 81.6% was reextracted in the two stage equipment.

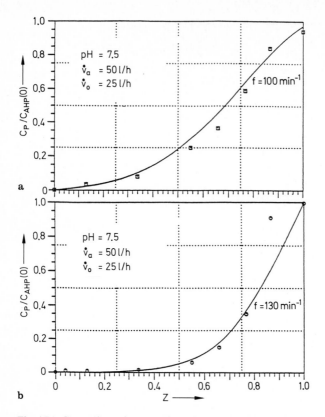

Fig. 4.76. Comparison of measured (*symbols*) and calculated (*curve*) longitudinal profiles of dimensionless penicillin concentration with regard to the complex concentration in the feed of the organic phase in a Kühni column at two different impeller speeds [332]

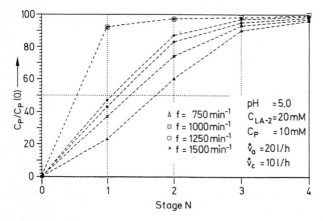

Fig. 4.77. Extraction of penicillin in four-stage mixer-settler. Dimensionless penicillin concentration with regard to the feed concentration as a function of the stage numbers at different impeller speeds [333] ◻ $C_{LA2} = 20$ mM; $\dot{V}_{aq} = 10$ l/h; pH 5 $C_P = 10$ mM; $\dot{V}_{org} = 10$ l/h

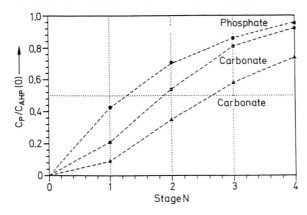

Fig. 4.78. Reextraction of penicillin in a four-stage mixer-settler. Dimensionless penicillin concentration in the aqueous phase with regard to the complex concentration in the feed of the organic phase with different buffers as function of the stage number [333]:

$\triangle$ carbonate buffer $\dot{V}_o = 15\,l\,h^{-1}$, $\dot{V}_a = 7.5\,l\,h^{-1}$; pH 7.2
$\square$ carbonate buffer $\dot{V}_o = \dot{V}_a = 10\,l\,h^{-1}$; pH 7.2
$\bullet$ phosphate buffer $\dot{V}_o = \dot{V}_a = 10\,l\,h^{-1}$; pH 7.5

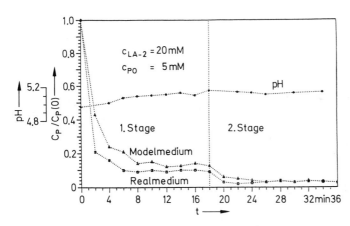

Fig. 4.79. Two-stage extraction; dimensionless penicillin concentration with regard to its feed concentration as a function of the time in centrifugal separator SA 01 of Westfalia [334]: ($\triangle$) model medium, ($\square$) fermentation broth and (*) pH

Penicillin was extracted at $C_P(0) = 20$ mM, $C_A(0) = 40$ mM, $V_a = 30$–$80\,l\,h^{-1}$, $V_o = 10$–$30\,l\,h^{-1}$, and pH 5 in a laboratory-scale system consisting of a centrifugal mixer, two single-stage extractors with built-in mixers TA-1 (Westfalia), and a centrifugal separator (Westfalia). At a phase throughput ratio of one, penicillin was completely extracted from the broth. With an increasing ratio of the aqueous phase throughput to that of the organic solvent, the extracted penicillin fraction was reduced, but its enrichment increased. For

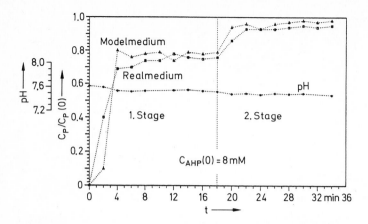

Fig. 4.80. Two-stage reextraction; dimensionless penicillin concentration with regard to the complex concentration in feed of the organic phase as a function of the time in centrifugal separator SA 01 [334]: ($\triangle$) model medium, ($\square$) fermentation broth and (*) pH

example, at a phase ratio of six, 61% is extracted, but penicillin is enriched by a factor of six. The reextraction was performed with $C_{AHP}(0) = 15\,mM$ and phosphate buffer at $V_a = V_o = 20\,l\,h^{-1}$ and pH = 7.5. 94% of the penicillin was reextracted under these conditions.

The pilot plant equipment consisted of three TA-7 units from Westfalia. The extraction was carried out with $C_P(0) = 10\,mM$, $C_A(0) = 20\,mM$, $V_a = 500\,l\,h^{-1}$, and $V_o = 120–460\,l\,h^{-1}$ at pH 5.1.

In Fig. 4.81, the residue fractions of penicillin $C_P/C_P(0)$ are shown after the first, second and third stages at different throughputs of the organic phase. At $V_o = 455\,l\,h^{-1}$ and after the third stage, penicillin is completely extracted from the broth.

In Fig. 4.82, the amount of reextracted penicillin after the first, second, and third stages are plotted. These results were obtained using two different buffers with $C_{AHP}(0) = 5.9\,mM$, $V_a = 140\,l\,h^{-1}$, $V_o = 240\,l\,h^{-1}$, and at pH 7.2–7.35. With phosphate buffer, more than 90% was reextracted, while 80% of the penicillin was reextracted with the less expensive carbonate buffer.

The throughputs of the reextraction are considerably lower than those of the extraction, since at high throughputs no satisfactory phase separation is possible. Thus, reextraction is the bottleneck of penicillin recovery.

It is recommended to carry out the extraction with higher aqueous-to-organic phase throughput ratios (e.g., at a ratio of 2.4 and corresponding penicillin enrichment of 2.4, 94% of the penicillin can be extracted) and the reextraction with a ratio of one, since at a high organic-to-aqueous phase throughput ratio, the penicillin residue fraction is high.

Extraction in Countercurrent Decanters [335, 339]. Penicillin was recovered from mycelium-containing fermentation broth by direct extraction with LA-2

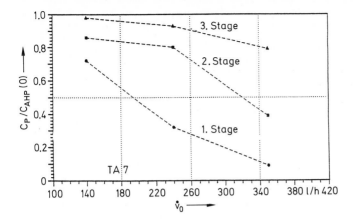

Fig. 4.81. Reextraction of penicillin; dimensionless penicillin concentration in the aqueous phase with regard to the complex concentration in the feed of the organic phase as a function of the throughput of the organic phase after the first, second and third TA-7 stages [334]: $\dot{V}_a = 140 \, l \, h^{-1}$ and $C_{AHP}(0) = 5.9 \, mM$

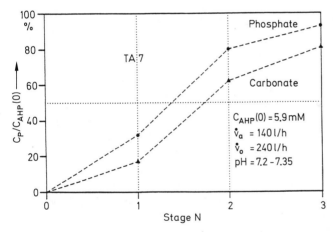

Fig. 4.82. Comparison of reextraction with carbonate and phosphate buffers in three TA-7-stage system as a function of the stage numbers [334]

and/or DITDA in *n*-butyl acetate in a countercurrent extraction decanter (Type CA 226-290, Westfalia Separator Co.) (Fig. 3.5, Chap. 3) at room temperature and steady state operation [317–319]. The operation conditions are given in Table 4.40.

In all runs, no phase separation problems occurred. The organic phase (extract) was clear and free of mycelium. In most of the cases, LA-2 was used as an amine extractant. In the first six runs, a fairly high amount of amine

Table 4.40. Operation conditions of the extraction of penicillin G with LA-2 and/or DITDA extractants in *n*-butyl acetate in counter current decanter [317]

No.	pH	V_a $1h^{-1}$	V_o $1h^{-1}$	V_a/V_o	C_P $g l^{-1}$	C_A $g l^{-1}$	$C_P/C_P(0)$ (%)
1	4.86	880	324	2.71	3.60	20	82.6
2	4.74	880	432	2.03	3.12	20	82.5
3	4.93	680	324	2.09	3.64	20	89.0
4	4.70	880	324	2.71	3.05	20	90.0**
5	4.63	880	650	1.35	3.14	20	94.1**
6	4.60	880	864	1.01	3.18	20	94.3**
7	5.0	600	200	3.00	3.60	15	76.0
8	5.0	820	220	3.70	3.90	15.6	72.0
9	5.0	820	360	2.30	3.90	15.6	87.0
10	5.05	750	500	1.50	15.0	60	96.0
11	5.00	660	245	2.70	3.50	14.2	80.0
12	5.00	660	245	2.70	3.50	7.2*	80.0
13	5.07	520	205	2.53	40.90	84.0*	85.0
14[a]	5.10	620	300	2.06	4.27	7.6*	70.0
15[a]	5.10	830	420	1.97	30.6	62.5*	92.0
16	4.95	620	330	1.88	4.28	0.0	19.0
17	5.1	690	310	2.22	30.2	0.0	17.0
18	2.3	690	310	2.22	30.2	0.0	61.0

* DITDA
** 50% of the organic phase was contacted with the aqueous phase in the static mixer before the extraction in the decanter was performed. This increased the contact time of the phase by 50%
[a] pH was set by phospheric acid

extractant was used (LA-2-to-penicillin ratios of five to six. The ratio of phase throughputs was varied between 2.71 and one. With decreasing phase through-put ratio, the extracted penicillin fraction increased from 0.82 to 0.94. The initial pH changed only slightly. In runs 7, 8, and 9, the amine-to-penicillin ratio was reduced to four, the pH was increased to 5, and the V_a/V_o ratio was increased to 3.6–3.9. The increase of this throughput ratio mainly caused a reduction of the fraction of extracted penicillin.

At high penicillin concentrations ($C_P(0) = 15 \, g l^{-1}$ and $C_A(0) = 60 \, g l^{-1}$, an amine-to-penicillin ratio of 4.0, and a low phase throughput ratio of 1.5, 96% of the penicillin was extracted (run 10).

For comparison, DITDA was also used as extractant (runs 11 and 12). With half the amount of DITDA, the penicillin fraction extracted is the same for both carriers.

In Fig. 4.83, the pH and the penicillin concentration in the organic phase and the dimensionless penicillin concentration in the aqueous phase are plotted as a function of the extraction time. One can observe that steady state had already been attained after 1–2 minutes. This behavior is typical for runs 1 to 6. The reduction of the throughputs of the phases (runs 7, 8 and, 9) results in an increase in the time required to attain steady state (Fig. 4.84).

Since buffers are expensive, their replacement by inexpensive chemicals for pH control would be desirable. In runs 14 and 15, the pH was set by a phos-

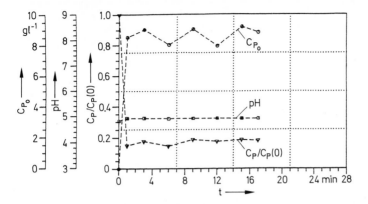

Fig. 4.83. Extraction of penicillin from the mold containing broth with LA-2-*n*-butylacetate in a counter current extraction decanter [335]. The pH value □ and the penicillin concentration in the broth with regard to the feed concentration $C_P/C_P(0)$ ▽ and penicillin concentration in the organic phase. C_{P_o} ○ are shown as a function of the extraction time [335]: pH 4.86; $\dot{V}_a = 880\,1\,h^{-1}$, $\dot{V}_o = 324\,1\,h^{-1}$; $C_P = 3.6\,g\,l^{-1}$, $C_A = 20\,g\,l^{-1}$

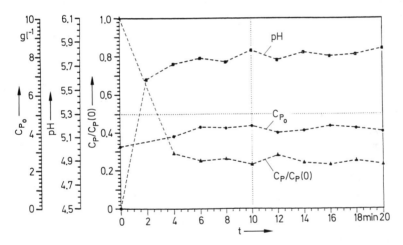

Fig. 4.84. Extraction of penicillin from the mold containing broth with LA-2-*n*-butylacetate in a counter current extraction decanter [335]: pH 4.7; $\dot{V}_a = 880\,1\,h^{-1}$, $\dot{V}_o = 324\,1\,h^{-1}$; $C_P = 3.05\,g\,l^{-1}$, $C_A = 20\,g\,l^{-1}$ C_{P_o} ○; pH □; $C_P/C_P(0)$ △

phoric acid solution. In order to avoid the decomposition of penicillin due to locally high acid concentrations, the acid was fed at first into the organic phase (run 14). However, this mode of operation reduced the efficiency of the decanter. Only 70% of the penicillin was extracted. Therefore, the phosphoric acid was fed into the broth (run 15). At a phase throughput ratio of about 2.5, 92% of the penicillin was extracted. At the same time, a penicillin concentration of 75% was obtained in the organic phase.

In order to compare the results with those in the absence of amine extractants, the measurements were repeated at a low penicillin concentration and pH
4.91 (run 16) and at a high penicillin concentration and pH 5.1 (run 17). Only
19.0 and 17.0% of the penicillin was extracted, respectively. By shifting the pH
from 5.1 to 2.3, the extracted penicillin was increased to 61%. However, 6.5% of
the penicillin decomposed during the short extraction time. At pH 5, no
penicillin loss could be detected.

The same solvent mixture (DITDA in n-butyl acetate) was used seven times
for extraction (at pH 5) and reextraction (at pH 9). The extraction performance
remained constant. This indicates that the extraction efficiency of the organic
phase remains constant: neither an enrichment of the coextracted compounds
nor a reduction of the amine concentration in the organic phase occurs during
the repeated extraction/reextraction process. This was also proven by the
determination of the DITDA concentrations in the organic phase during these
investigations.

Since the contact time of the phases influences the extraction performance,
the contact time was increased by 50% using a motionless mixer. However, the
amount of extracted penicillin increased by only 4%.

Since the drum rotation speed is high (4350 min^{-1}), one would expect some
damage of the mycelium due to the mechanical stress in the extraction decanter.
The hyphae were, therefore, examined under a microscope before and after the
extraction process. It was found that, during the extraction, the hyphae became
shorter and the protein content in the broth increased from 0.2–0.8 gl^{-1} to
1.2–1.6 gl^{-1}, i.e., by 50 to 100%. However, this protein content increase did not
influence the phase separation.

Based on these investigations, direct extraction of the mycelium containing
broth with amine extractants in a countercurrent extraction decanter can be
recommended.

Extraction in Three-Phase Systems [336, 339]. In the preceeding chapters,
extraction and reextraction of penicillin were considered as two separate processes. In liquid membrane systems [351, 352], these two processes are carried
out simultaneously. Therefore, investigations were carried out with a special
three-phase system (Fig. 4.85) in which two aqueous phases with different pH
values were separated by an organic solvent phase that contained an amine-
extractant acting as a carrier for penicillin. With this carrier, penicillin was
transported from the aqueous phase with the lower pH (phase I) to the one with
the higher pH value (phase II). In these investigations, n-butyl acetate was used
as the organic solvent phase and LA-2, which is only soluble in the organic
solvent phase, was used as carrier. At the interface of the acidic phase I (pH 5.5)
with the solvent phase, penicillin acid anions form a penicillin-amine complex
with LA-2, while at the interface of the alkaline phase II (pH 7.5) with the solvent
phase, the complex dissociates and penicillin acid anions are released. The amine
carrier permeates back to the phase I interface, where it reacts with the penicillin
acid anion again. By using buffers in both of the aqueous phases, a concentra-

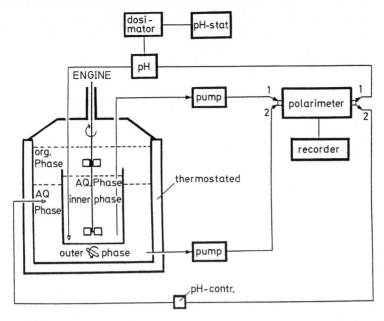

Fig. 4.85. Schematic arrangement of the three-phase reactor for the investigations of the simultaneous extraction of penicillin G by LA-2-*n*-butylacetate at pH 5.5 and its reextraction with buffer solution at pH 7.5 [336]

tion gradient of the complex is maintained between phases I and II, and a concentration gradient of the carrier between phases II and I. In this way, penicillin can accumulate in phase II.

The kinetics of this extraction-reextraction process was investigated in this three-phase system, and it was found that the following simple consecutive reaction scheme could describe the kinetics:

$$P_{aI}^- + H_{aI}^+ \xrightarrow{k_I} AHP_o \underset{k_{III}}{\overset{k_{II}}{\rightleftharpoons}} A_{oII} + P_{aII}^- + H_{aII}^+ \tag{4.84a}$$

$$A_{oII} \xrightarrow{k_{IV}} A_{oI}, \tag{4.84b}$$

where P_{aI}^- and P_{aII}^-, the penicillin anion, H_{aI}^+, and H_{aII}^+, the proton concentrations in the aqueous phase I and II, A_{oI} and A_{oII} are the amine (carrier) concentrations at the interfaces of the organic membrane phase with phases I and II, and A_{AHPo} is the concentration of the complex in the organic membrane phase.

At LA-2 concentrations in the organic solvent phase above 15 mM, the complex formation is independent of the carrier concentration, and depends only on the penicillin concentration. Therefore, the reaction scheme can be simplified to

$$P_{aI} \xrightarrow{k_I} P_o \underset{k_{III}}{\overset{k_{II}}{\rightleftharpoons}} P_{aII}. \tag{4.85}$$

The complex formation and dissociation can be described by the following 1st order reactions:

$$-\frac{dC_{PaI}}{dt} = k_I C_{PaI},\tag{4.86a}$$

$$\frac{dC_{Po}}{dt} = k_I C_{PaI} - k_{II} C_{Po} + k_{III} C_{PaII},\tag{4.86b}$$

$$\frac{dC_{PaII}}{dt} = k_{II} C_{Po} - k_{III} C_{PaII}.\tag{4.86c}$$

By measuring the concentrations of the penicillin in each of the three phases, the model parameter k_I, k_{II}, and k_{III} were identified. It is possible to describe the course of the penicillin concentration in the three phases (Fig. 4.86). The difference between the calculated and measured courses are due to the coextraction of the buffers and the decomposition of penicillin during the runs.

Due to the small specific interfacial area of the phases I and II and the large volume of the organic solvent phase, the extraction process is very slow. Penicillin temporarily accumulates in the organic phase. One can, therefore, conclude that the extraction of penicillin from phase I is influenced considerably by its reextraction into phase II.

The model parameters are functions of several factors, including the pH in the aqueous phases and mass transfer through the organic solvent phase. If these functions could be determined, then the extraction rate of penicillin can be calculated.

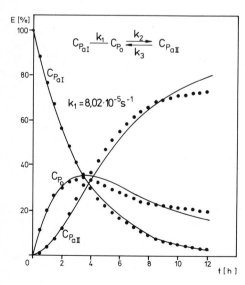

Fig. 4.86. Variation of the dimensionless penicillin concentration with regard to its initial value in the phase I (C_{PaI}), in the organic phase (C_{Po}) and in the aqueous phase II (C_{PaII}) with the extraction time. Comparison of measured (*symbols*) and with Eq. (4.85) calculated (*curve*) values [336]

Extraction/Bioconversion Process [341–343]. Liquid membranes are usually applied for the selective extraction of a medium component [349–353]. However, the extraction can be combined with the conversion of this component. The combination of a selective extraction of a component with its selective conversion is especially interesting.

Scheper et al. [342] and Barenschee et al. [354] reported on the selective extraction of penicillins G and V and their enzymatic conversion to 6-APA. The principle of the process is shown in Fig. 4.87. Penicillin acid anion (P^-) and proton (H^+) form a complex with the carrier (C) at the interface I (outer aqueous phase/organic solvent (membrane) phase). The complex (CHP) permeates through the membrane phase and decomposes at the interface II (membrane phase/inner aqueous phase) into P^-, H^+, and C. Penicillin acid anions are enzymatically split into 6-amino penicillic acid (6-APA) and phenyl acetic acid anion (PhA^-) in the inner aqueous phase. PhA^- forms a complex with the proton and the carrier at interface II. The complex (CHPhA) permeates through the membrane and dissociates at interface I into PhA^-, H^+ and C. The phenyl acetic acid anion returns to the outer phase. The carrier forms a new complex with penicillin acid anion and proton, and the process continues. The other product, zwitterionic 6-APA, cannot pass through the membrane phase and, therefore, accumulates in the inner aqueous phase. The extractions were carried out at different penicillin concentrations and at pH 6 in the outer phase, with LA-2 (carrier) and Span 80 (surfactant) in kerosene as the membrane phase, and the reaction (catalyzed by penicillin G-amidase enzyme at pH 8 occurring in the inner phase.

The best results were obtained with a membrane phase consisting of 7.5% Span 80 and 1.5% LA-2 in kerosene. Under these conditions 6-APA is completely retained.

Laboratory-scale extractions were carried out in a continuously-operated Kühni column (Fig. 4.53). The outer phase and the emulsion phase were separated immediately in a settler. The outer phase (with the reextracted phenyl acetic acid) was recycled to the fermentor, where PhA is used by the mold as precursor for penicillin formation. The emulsion was broken in a continuously

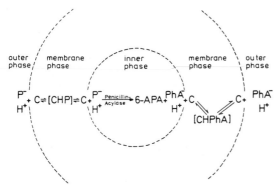

Fig. **4.87.** Reaction scheme of penicillin conversion to 6-APA by means of penicillin G-amidase immobilised in a liquid membrane emulsion [343]

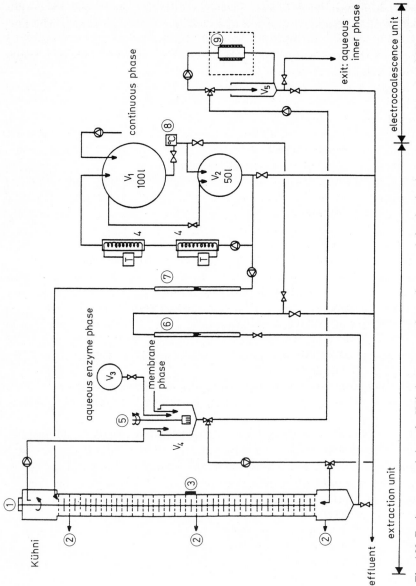

Fig. 4.88. Equipment consisting of a Kühni extractor and a coalscence unit for the simultaneous extraction of penicillin and its enzymatic conversion to 6-APA [343]

operated electrocoalescence unit. The membrane phase was recycled and reused for emulsion formation (Fig. 4.88). The 6-APA was converted with d-phenylglycine into ampicillin at pH 6 by the same enzyme. The enzyme was recovered with an ultrafiltration membrane, recycled, and immobilized in the liquid membrane again.

4.6.1 Symbols for Sect. 4.6

a_p	specific interfacial area
A_o	amine in the organic phase
AHP_o	penicillin-amine complex in the organic phase
AHX	coextracted ion-amine complex in the organic phase
C_A	concentration of A_o
C_{A0}	feed concentration of A_o
C_{At}	$C_A + C_{AHP}$
$C_A(0)$	concentration of A_o in feed or its initial concentration
C_{AHP}	concentration of AHP_o
C_{AHP0}	feed concentration of AHP_o
$C_{AHP}(0)$	concentration of AHP in feed or its initial concentration
C_{AHX}	concentration of AHX
C_H	concentration of H_a^+
C_{HPa}	concentration of HP_a
C_{HPo}	concentration of HP_o
C_{Pa}	concentration of P_a^-
C_{Pt}	$C_P + C_{HPa} + C_{HPo} + C_{AHP}$
C_{P0}	feed concentration of P_a^-
$C_P(0)$	concentration of P_a^- in feed or its initial concentration
C_X	concentration of X^- anion
Cl_a^-	chloride anion in the aqueous phase
C_{QCl}	concentration of $Q^+Cl_o^-$
C_{QClt}	$C_{QCl} + C_{QP}$
C_{QP}	concentration of $Q^+P_o^-$
C_{Xt}	$C_X + C_{HXa} + C_{HXo} + C_{AHX}$
d_{32}	Sauter droplet diameter
E	degree of extraction
E_R	degree of reextraction
f	stroke frequency, impeller speed
H_a^+	proton in the aqueous phase
j_{Ao}	fux of A_o
j_{AHPo}	flux of AHP_o
j_{HPa}	flux of HP_a
j_{HPo}	flux of HP_o
j_{Pa}	flux of P_a^-
K	distribution coefficient (Eq. 4.61)

K_G equilibrium constant (Eq. 4.63)
K_p distribution coefficient (physical extraction)
K_X equilbrium constant (Eq. 4.70)
k_A mass transfer coefficient
k_{AHP} mass transfer coefficient of AHP_o
k_{HPa} mass transfer coefficient of HP_a
k_{HPo} mass transfer coefficient of HP_o
k_P mass transfer coefficient of P_a^-
N number of stages in the cascade model
P_a^- penicillin acid anion in the aqueous phase
P_c partition coefficient
$Q^+Cl_o^-$ quaternary ammonium chloride
$Q^+P_o^-$ penicillin acid anion quaternary ammonium salt
X_a^- coextracted anion
V_a throughput of the aqueous phase
V_o throughput of the organic phase
ε holdup of the organic phase
τ mean residence time in a stage of the cascade model

4.7 Integrated Processes

Recovery and purification of biotechnological products consist of several sequences of unit operations. If the product is secreted into the culture medium, the recovery starts with separation of the cells by filtration, membrane separation or centrifugal decanting. Volatile and heat persistent products are often recovered from clarified culture medium with distillation. Nonvolatile, heat sensitive products are recovered from the clarified culture medium by adsorption or extraction. Purification is performed by a series of extraction or adsorption/elution processes. Several antibiotics are recovered from the clarified culture medium by adsorption by ion exchange resins followed by elution with aqueous salt solution (e.g. clavulanic acid). Sometimes the adsorption/elution process is repeated many times (e.g. in the case of Carbapenem compounds eight times).

Several biotechnological production processes are impaired by product inhibition. This holds especially true for batch and fed-batch processes in which the product concentration attains high values at the end of the process. By integrated product formation-recovery the product concentration can be kept at a low value and the productivity can be increased considerably.

Adsorption [355–357], distillation [358], precipitation [359], electrophoresis [360, 361], pervaporation [362–365], gas stripping [366], dialysis [367] reverse osmosis [368], and also extraction can be applied for integrated product recovery. For volatile products pervaporation and for non volatile products the adsorption and extraction are the most common methods.

The bottleneck of extraction is the availability of biocompatible non toxic inexpensive solvents with high distribution coefficients.

Most of the investigations dealt with the extraction of primary metabolites, like ethanol, acetone-butanol, carboxylic acids from the fermentation broth during their formation. The question of biocompatibility and distribution coefficients of the carbon bonded oxygen bearing extractants are already considered in Chapters 4.1 and 4.2. Here the results are summarized and new trends are presented.

Ethanol Recovery. After systematic investigation of a large number of solvents, oleyl alcohol was identified as the best biocompatible solvent for ethanol [369–373]. Daugulis et al. [372] investigated the economy of ethanol extraction. They found that conventional and extractive ethanol production plants have almost identical design. It should be possible to retrofit a conventional plant with an extractive unit a substantial cost saving over the grass roots plant. The optimal conditions of operation were evaluated by a computer programme. Three different strategies were investigated. For the first two, operating conditions were based on the optimal conditions (medium dilution rate $D = 0.15 \, h^{-1}$ and solvent dilution rate $D_S = 1.25 \, h^{-1}$ and $D = 0.18 \, h^{-1}$, $D_S = 1.50 \, h^{-1}$ respectively). They provided ethanol selling prices of 32.70 c/l and 31.01 c/l respectively, compared to 45.06 c/l for conventional plant. The third strategy fitted the capacity of the extraction unit better to the product formation unit. This caused an increase of D_S to $2.0 \, h^{-1}$. This yielded an ethanol selling price of 29.44 c/l. However, as expected, the largest component (60%) of the overall production cost of ethanol continues to be the raw material cost (mainly substrate cost) followed by utilities (largely energy) at 26%. Equipment depreciation accounts for only 3% of the production costs of ethanol indicating the relative unimportance of numerous efforts aimed at developing better novel reactors for ethanol production process. Relative to the conventional ethanol production, the grass roots extractive unit has a major saving in energy, because it uses significantly higher substrate concentration, i.e. less water [372].

When using immobilized cells, the toxicity of the solvent for the cells can be reduced, if protective agents (Poropack Q, castor oil etc.) are used [374–377]. Ethanol was also extracted by oleyl acid from the broth of immobilized cell culture [378].

However, cell immobilization has several problems: low mechanical stability, leakage of viable cells, not negligible internal mass transfer resistance. Therefore, its practical application in large scale depends on improved immobilization matrix properties.

The extractant oleyl acid esterified with ethanol by lipase (from *Mucor miehei*), which was coimmobilized with the cells (*Saccharomyces baynus*) in microemulsions in presence of phosphatidylcholine, yielded a tenfold increased distribution coefficient for ethanol [379].

Other authors applied membrane extraction using hollow fiber module and different extractants (e.g. 2 ethyl-1-hexanol, ethylene or propylene glycol, isotridecanol) [380–384]. The ethanol production with membrane extraction is at least as competitive as the conventional process (the ethanol costs can be reduced by 3–4%).

Also at low distribution coefficient the ethanol recovery is efficient, if ethanol is continuously removed from the extractant by distillation. This combination of solvent extraction and extractive distillation was performed with tri-n-butyl phosphate/Isopar M extractant [385].

To avoid the direct contact of the culture medium with the solvent, they are separated by a thin gas layer in the pores of a membrane module. This separation is called transmembrane distillation [381, 386]. However, this recovery is inefficient.

Another method was recommended by Christen [382] to separate the cultivation medium from the solvent during the extraction of ethanol. A teflon sheet was soaked with isotridecyl alcohol and used as solid-supported liquid membrane (SLM). This assembly permitted biocompatibility, permeation efficiency, and stability. By the removal of ethanol from the cultures, the ethanol productivity was increased by a factor of 2.5, and the cell viabilty was maintained above 95%. At the same time, ethanol was purified from the medium components. The ethanol was removed from the extractant by aqueous and gaseous stripping, respectively. It seems that the former has a higher performance.

All of these methods are at a laboratory stage. It is not yet clear, if they can be realized in practice.

Acetone-Butanol. Similar to the ethanol formation, the production of acetone-butanol by *Clostridium acetobutylicum* is impaired by product inhibition already at 2% product concentration, where acetone inhibition can be neglected in comparison to that of butanol inhibition. Pervaporation, adsorption and extraction are the common techniques for product removal during the product formation. As extractants n-decanol [387], dibutylphthalate [388], polypropylene glycol [389] and oleyl alcohol [390, 391] were used.

By extraction of the cell-free medium with n-decanol the productivity was increased fourfold in continuous culture [387]. By extraction of the acetone-butanol from cell-containing medium with dibutylphthalate as extractant the product concentration was increased from 18 to 20 g l^{-1} to 28 to 30 g l^{-1} [388].

With oleyl alcohol extraction in a Karr column, the productivity was increased by 20% in batch culture, which lowered the capital costs by 20% [390]. In continuous culture the productivity was around 70% higher than in the batch culture [392].

With extraction of acetone-butanol by oleyl alcohol across silicon tubing, the productivity was increased fourfold [391], and using isopropyl myristate the productivity was doubled [393], both of them in batch culture.

Clostridium acetobutylicum was immobilized in the interfiber space of a microporous hollow fiber membrane module (Cellgard X-20, Hoechst Celanese). The culture medium was pumped through the interfiber space, and the product was extracted with 2-hexyl-1 hexanol by pumping it through the intrafiber bores. The re-extraction was performed in a second module. The productivity was increased by 40% [394].

Butanol/isopropanol was removed from the cultivation medium of immobilized *Clostridium isopropylicum* by means of pervaporation across a supported liquid membrane consisting of oleyl alcohol liquid membrane phase. The butanol concentration in the permeate was 230 kg m^{-3}, which is 50 times higher than in the culture medium [395].

Groot et al. [396] compared the economy of different technologies for integrated butanol recovery: stripping, adsorption, liquid-liquid extraction, pervaporation and membrane solvent extraction [396]. From these, pervaporation and liquid-liquid extraction was found to have the greatest potentials.

The calculations are based on the extraction of 75% butanol in a Rotating Disc Contactor with oleyl alcohol and 50/50 oleyl alcohol/decane mixture as extractant. In Table 4.41 the energy requirements in the integrated processes are compared.

The total heat of recovery with extraction is indeed higher than with pervaporation, but the liquid-liquid extraction has a higher selectivity of alcohol/water separation and makes it possible to carry out the separation inside the fermenter.

Carboxyl Acid Recovery. Similar to alcohols the cell growth and product formation are reduced at high product concentrations. This can have two effects: 1) at high acid concentration the pH value changes to low values. 2) if the pH value is kept constant, high acid concentration has a detrimental effect on the cell biology. At low pH this effect is increased, because of higher fraction of acids in the non-dissociated form. Only non-dissociated acids can pass the cell membrane. Therefore, several authors dealt with the integrated acid production recovery. Four approaches can be distinguished:

- the extractant contacted with the immobilized cells [397–401],
- separation of the cells from the extractant by a membrane [402, 405, 406],
- separation of the cells from the extractant by a composite membrane [403], and

Table 4.41. Energy requirements in integrated processes [396]

Recovery method	Product	Estimated total heat of recovery (MJ/kg ABE)
Stripping	B, and AE mixture	21
Adsorption	B, and AE mixture	33
Extraction and perstraction[1]	ABE mixture	14
Pervaporation	B, and AE mixture	9

A, acetone; B, butanol; E, ethanol.
[1] Oleyl alcohol as extractant.

– conversion of the acids into a hydrophobic derivative of high solubility in the extractant [404].

Yabannavar and Wang produced lactate by the immobilized *Lactobacillus delbrückii*, and the acid was extracted at low pH by TOPO + dodecane in Isopar M [397] and Alamine 366 in oleyl alcohol [400], respectively. The latter increased the productivity from 7 to 12 $g l^{-1}$ (gel) h^{-1}. They also used soybean oil as supplement to the Kappa-Karragenaan matrix to trap the diffusion solvent molecules [399] and simulated the process [398]. This indicated that a significant build-up of the inhibitory product occurs in the beads.

Bar and Gainer [401] produced lactic acid by immobilized *Lactobacillus delbrückii*, citric acid by immobilized *Aspergillus niger*, and acetic acid by immobilized *Acetobacter aceti*. The acids were extracted with different extractants (long-chain hydrocarbons, perfluoro decalin, methyl oleate). The main problem besides the toxicity of the extractants is the low pH necessary for the acid extraction.

Bartels et al. [402] used microporous membrane modules to extract lactic acid by Alamine 336 + MIBK.

Siebold [405] produced D-lactic acid by *Lactobacillus delbrückii* DSM 20072 on MRS medium (Casein, meat extract, yeast extract, nutrient salts and glucose), and L-lactic acid by *Lactobacillus casei* ssp rhamnosus DSM 20021 on MRS medium and by *Lactobacillus salivarius* ssp salivarius on MRS medium in batch and continuous culture. The highest productivity was obtained with *L. salivarius*.

The product extraction was performed by kerosene and butylacetate with Amberlite LA2 (secondary amine), Hoey F2562 (secondary amine) as well as Hostarex A327 (tertiary amine) and Alamine 336 (tertiary amine) in presence of different modifiers: long-chain alcohols (isodecanol), alkylphosphates (tributylphosphate) and acidic organic compounds (4-nonylphenol), to improve the solubility of the acid-carrier complex. The highest degree of extraction was obtained with Hostarex A327.

When using Cyanex 923 (mixture of trioctyl- and trihexylphosphinoxides) no modifier was necessary. Since the lactic acid production by *L. casei* and *salivarius* are enhanced in presence of acetate and citrate, the medium was supplemented by these acids. Kerosene was biocompatible, 4-nonlyphenol was not compatible at all, all others were only partly compatible. The biocompatibility of Cyanex 923 is better (20% extractant reduced the lactate concentration by 30%) than that of Hostarex A327 (20% extractant reduced the lactate concentration by 45%). With 10% TBP or isodecanol the product concentration is reduced by 20% [406].

Prasad and Sirkar [403] used microporous hydrophilic and composite membranes to extract succinic acid and acetic acid from aqueous solutions by *n*-butanol and MIBK respectively. However, no data on the biocompatibility of this system were published. In order to increase the solubility of the acids in extractants, they were converted into hydrophobic esters by lipase. The esterification with oleyl alcohol increased the apparent distribution coefficients in CBO

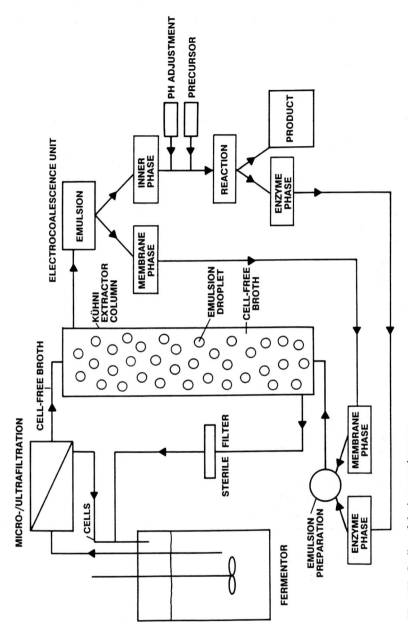

Fig. 4.89. Outline of the integrated process

extractants four to 15-fold [404]; however, the back extraction becomes more difficult. It is doubtful whether the economy of the acid recovery is improved by this technique.

Recovery of Secondary Metabolites. Only few investigations have been published on the integrated production-extraction of secondary metabolites. England et al. [407] investigated the extraction of codein/norcodein with functionalised polysiloxanes and Mavituna et al. [408] capsaicin with sunflower oil. Cycloheximide and gibbelinic acid were extracted from the permeate by substituted polyglycols (Genapol 2822 and Nonlyphenol-8-1-polyethylene, Bayer AG. Leverkusen) by Onken and coworker [409, 410].

The extraction of penicillin G from the filtered cultivation medium of *Penicillium chrysogenum* and its conversion into 6-amino penicillanic acid (6-APA) and phenyl acetic acid (PhA) at pH 8 was performed in a 10-l Kühni extractor during the production by means of penicillin G-amidase immobilized in an emulsion liquid membrane (ELM) Amberlite LA-2-carrier system. 6-APA was enriched in ELM, the PhA was returned into the cultivation medium. After electrocoalescence of ELM, the 6-APA was converted into ampicillin with the same enzyme at pH 6, the liquid membrane phase and the enzyme were recycled are reused [411].

The advantages of this system are that the penicillin concentration can be kept at a low level ($7 \, g \, l^{-1}$) and the penicillin losses during production can be reduced by a factor of three. On the other hand, by recycling of PhA, the precursor (PhA) consumption can be considerably reduced. The process stages of ampicillin production can be reduced by a factor of three. Figure 4.89 shows an outline of the integrated system.

The disadvantage is that the stability of the liquid membrane system is reduced at high penicillin concentrations. Therefore liquid membrane extraction into the organic phase (Amberlite LA 2 in Kerosene) and back-extraction (into the aqueous enzyme solution) were carried out in a membrane module, in which the phases were separated with a very thin Cellgard membrane [412]. Because of the rather high mass transfer resistance across the solid membrane, the performance level of this system is lower than that of the emulsion liquid membrane system [412].

Integrated product formation/product recovery by extraction will gain in importance in the future.

4.8 Separation of Complex Mixtures

When metabolites are produced with highly mutated strains or recombinant strains optimized by metabolic design and under optimal cultivation conditions, the product concentrations attain high levels (e.g., penicillin, lysine). The recovery and purification of these products is relatively easy: with a single recovery procedure and after a few purification steps, sufficient product quality can be attained.

However, several purification steps are necessary if the product concentration is low and the accompanying compounds are chemically similar, e.g., if particular amino acids are to be recovered from protein hydrolysates.

When the components to be separated have only a single charged group and their pK_a-values are different, their separation is possible by means of suitable extractants and at an appropriate pH value, e.g., the separation of formic acid ($pK_a = 3.7$) from acetic acid ($pK_a = 4.76$). However, the separation of the higher carboxylic acids based on their pK_a-values is not possible (the pK_a-values of propanoic acid, butyric acid, pentanoic acid, hexanoic acid, and heptanoic acid are in the range of 4.85–5.0). The pK_a-difference between propanoic acid (4.85) and lactic (3.86) and pyruvic acid (2.49) is large enough for a separation. Also maleic acid ($pK_a = 1.93$, 6.14) and fumaric acid ($pK_a = 3.02$, 4.38) (cis- and trans-isomers) can be separated by ion-pair extraction.

If the compounds to be separated have two or more charged groups with negative and/or positive charges, they form zwitter ions in a neutral range. Typical examples area amino acids with at least two pK-values (one is characteristic for the anion and the other for the cation), which are in a relatively narrow range ($pK_1 = 2$–3 and $pK_2 = 9$–10) and, therefore, they are not suited for separation.

On account of the zwitterionic character of the amino acids, the carbon-bonded, oxygen-containing extractants are not well-suited for their separation and recovery. Phosphorus-bonded, oxygen-containing (PBO) extractants form associate-complexes with amino acids, which are hydrated with one or two water molecules. PBO extractants are more appropriate for amino acid recovery. Properties of the side groups (polar, apolar, positive charge, negative charge) can be used for their separation. In the following, some examples are shown.

The separation of lactic acid ($pK_a = 3.86$) from amino acids ($pK_1 = 2$ to 3) is possible. When the aqueous solution of lactic acid and the amino acids are extracted with 40% HOSTAREX A 327 (1:1 tri-n-octyl/tri-n-decylamine-mixture) in presence of 10% isodecanol in 50% kerosene at pH 1.8, the lactic acid is transferred into the organic phase, and the amino acids remain in the aqueous phase. After re-extraction with 1 M NaOH solution, pure lactate solution is obtained [415]. Similar results can be attained with TOPO (trioctyl-phosphinoxide) and Amberlite LA-2 [415] (Fig. 4.90).

The separation of particular carboxylic acids is much more difficult.

In lactic acid production with several *Lactobacilli* the supplementation of the medium with acetic and citric acids increases the productivity [405]. However, lactic acid recovery and purification means its separation from the other carboxylic acids.

In Fig. 4.4, the degree of coextractions of acetic, lactic and citric acids with 30 wt% Hostarex A 327 and 10 wt% isodecanol at different temperatures are shown. Acetic acid has the highest, and lactic acid the lowest extraction degree.

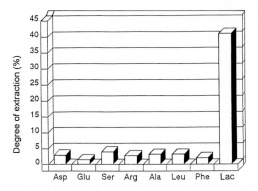

Fig. 4.90. Separation of lactic acid from amino acids with 40% HOSTAREX A 327 + 10% ISODECANOL + 50% KEROSENE at pH 1.8 and back extraction with 1 N NaOH [415].

However, the difference is rather low. In a multistage extractor their separation would be possible.

The degree of reextraction of lactate with 0.1 N NaOH is high (90%). The degree of reextraction with water amounts to only 55%. The reextraction of free acids from the Hostarex A 327 and isodecanol containing organic phase by 2 M HCl yields the highest degree of reextraction of citric acid, a high degree of extraction of lactic acid and a low degree of extraction of acetic acid (Fig. 4.91). A separation of acetic acid is possible in a multistage extractor. However, the total yield of the lactic acid is rather low (45%). The total yield with 1 molar H_2SO_4 is somewhat lower.

The degrees of coextraction of the same acids with 30 wt% Cyanex 923 at different temperatures are shown in Fig. 4.5. Again the extraction degree of acetic acid is the highest, and that of the citric acid the lowest. Above 40 °C no citric acid is extracted. Its separation is possible in a single stage equipment. The separation of lactic and acetic acids is only possible in a multistage extractor.

The reextraction of lactate and acetate with NaOH is not possible, because a stable emulsion is formed. The equilibrium lactic acid concentration in the

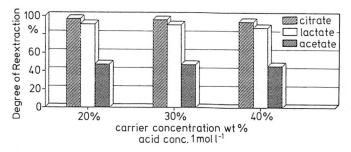

Fig. 4.91. Selectivities of reextraction of citric, lactic, and acetic acids from Hostarex A 327 system [406].

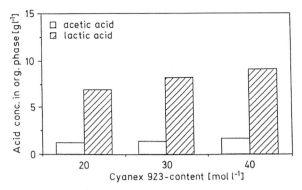

Fig. 4.92. Equilibrium lactic acid and acetic concentrations in the organic phase with different concentrations of Cyanex 923 [406].

organic phase with Cyanex 928 amounts to about 8% and the equilibrium acetic concentration to about 2.5–3.0% (Figure 4.92) [406].

The reextraction of the free acids with 2 M HCl yields the highest degree of reextraction of lactic acid, and low degrees of reextraction of acetic and citric acids (Fig. 4.93). Thus, the separation of the three acids is possible, but the total yield of the lactic acid is still low (max. 20%).

If the ultrafiltered cultivation medium of *Lactobacillus salivarius* spp. salivarius is extracted with different amounts of Hostarex A 327 and 10% isodecanol, about a 50% degree of extraction of acetic and lactic acids, respectively, can be obtained (Fig. 4.94). The equilibrium lactic acid concentration in the organic phase amounts to about 10% and because of the low acetate content of the cultivation medium, the equilibrium acetic acid concentration in the organic phase is low (2.5%).

With Cyanex 923 a lower degree of extraction of acetate and lactate, respectively, is obtained from the ultrafiltered cultivation medium (Fig. 4.95). The equilibrium lactic acid concentration in the organic phase amounts to about 8% and the equilibrium acetic acid concentration to about 2.5–3.0%.

The reextraction of lactic acid with 2 M HCl from the carrier containing organic phase yields about a 75% degree of extraction with Hostarex A 327, and

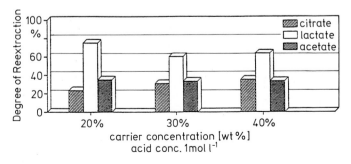

Fig. 4.93. Selectivities of citric, lactic, and acetic acids from a Cyanex 923 system [406].

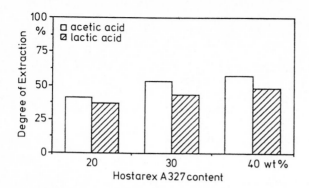

Fig. 4.94. Extraction degrees of acetic and lactic acids with different concentrations of Hostarex A 327 [406].

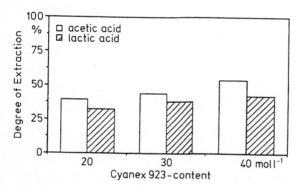

Fig. 4.95. Extraction degrees of acetic and lactic acids with different concentrations of Cyanex 923 [406].

55% with Cyanex 923 (Fig. 4.96). Again the total yield coefficient is low (37% for Hostarex and 22% for Cyanex) [406].

The separation of amino acids from each other is also a difficult task.

For the separation of isoleucine and leucine, which have apolar side groups, and lysine, which has a positive charge at pH 6, TOMAC (methylrioctyl-ammoniumchloride) and 6% decanol in methylcyclohexane were used at pH 10 to 11. As expected, the basic amino acid (lysine) remains in the aqueous phase and the amino acids with apolar, noncharged side groups (leucine and isoleucine) are transferred into the organic phase. After their reextraction with HCl or NaCl solutions in the aqueous phase, leucine and isoleucine can be separated by an additional extraction by means of an anion exchanger in cyclohexane. Leucine is transferred into the organic phase, and isoleucine remains in the aqueous phase [418].

According to an alternative method, the amino acids are extracted from a 4 M NaCl solution by tetra-dodecylammoniumbromide, an anion exchanger,

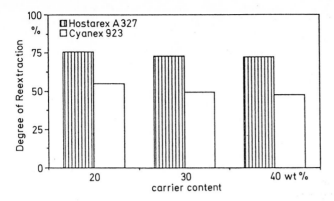

Fig. 4.96. Reextraction degrees of lactic acid from organic phase with different concentrations of Hostarex A 327 and Cyanex 923 [406].

in the presence of 6% decanol in xylene at pH 12. Leucine is again transferred into the organic phase, and isoleucine remains in the aqueous phase. A comparison of the degrees of extractions of 10 mM leucine and isoleucine by extraction with tetrahexyl-, tetraoctyl- and/or tetradodecylammoniumbromide in the presence of 2%, 4% and/or 6% dodecanol in xylene at pH 12.4 indicates that with the increasing length of the alkyl chain the degrees of extraction of both components increase, but the selectivity of leucine diminishes. Therefore, the best separation can be attained with tetrahexylammoniumbromide. On the other hand, the difference of the selectivities increases with increasing NaCl concentration [418].

Tryptophan, an amino acid with an apolar side group, but of high solubility in water, and tyrosine with a polar side group, but of low solubility in water, are not separated from the aqueous phase by extraction with 15 mM TOMAC in xylene at pH 10 to 11. Both of them are transferred into the organic phase. However, they can be separated by re-extraction with a 1.5 M NaCl solution. Tryptophan remains in the organic phase and tyrosine is transferred into the aqueous phase [413]. With increasing NaCl concentration, their extraction degrees increase. The difference of their selectivity attains its highest value at 0.15 M NaCl.

When extracting arginine and lysine, amino acids with a basic character, as well as aspartic acid with 0.3 M TOMAC in xylene at pH 12, the aspartic acid is transferred into the organic phase, as expected and arginine and lysine remain in the aqueous phase. The degree of extraction of aspartic acid as a function of the equilibrium pH-value passes a maximum at pH 12 due to the coextraction of the OH^--anions [414].

The following example from a patent [416] indictes the difficulty of amino acid separation from a complex mixture. (Table 4.42).

This extractive separation procedure uses the differences in the polarity of the side groups and the solubility of amino acids in water (at the separation of

Table 4.42. Separation of amino acids by extraction [416]

Amino acid solution (aqueous phase I) consisting of
A) amino acids with nonpolar side groups
    Ala*, Val, Leu, Ile, Pro*, Phe, Met
    (*in aqueous phase well-soluble)
B) amino acids with polar side groups
    Gly, Ser, Thr,
C) amino acids with basic side groups
    Arg, Lys, His
D) amino acids with acidic side groups
    Asp, Glu
in aqueous solution in presence of 150 g/l NaCl

Extracted with the extractant
E_1 (75% D2EHPA (di-2-ethylhexyl phosphoric acid, cation exchanger) in presence of
10% ethyl-2-hexanol in 15% kerosene at pH 2.7)

	The orgaic phase is extracted with 2 N HCl
Aqueous phase 1	Aqueous phase 2
A) Ala, Pro	A) Val, Leu, Ile, Phe, Met
B Gly, Ser, Thr	B) —
C) Arg, His, Lys	C) —
D) Asp, Glu	D) —
92 g/l NaCl	80 g/l NaCl

Extracted with E_1 at pH 7		Extracted with E_1 at pH 3	
Aqueous phase 1.1	Org. phase 1.2	Aqueous phase 2.1	Org. Phase 2.2
A) Ala, Pro	—	Val, Met	Leu, Ile, Phe
B) Gly, Ser, Thr	—	B) —	—
C) Lys, His	Arg	C) —	—
D) Glu, Asp	—	D) —	—
80 g/l Nacl		33 g/l NaCl	
	Reextracted with 2 N HCl into aq. phase 1.2.1		Reextracted with 2 N H$_2$SO$_4$ into aq. phase 2.2.1

Aqueous phase 1.1	Aqueous phase 2.2.1 142 g/l Na$_2$SO$_4$

Aq. phase 1.1 extracted with E_1 at pH 3	Aq. phase 2.2.1 extracted with E_2 (15% Aliquate 336 (tri (C$_8$-C$_{10}$), monomethyl ammonium chloride) and 85% Solvesso 150 at pH 10

Aq. phase 1.1.1	Org. phase 1.1.2	Aq. phase 2.2.1.1	Org. phase 2.2.1.2
A) Ala, Pro	—	A) Leu, Ile	Phe
B) Gly, Ser, Thr	—	B) —	—
C) Lys	His	C) —	—
D) Glu, Asp	—	D) —	—
		92 g/l Na$_2$SO$_4$	
	Reextract. with 2 N HCl in the aq. phase 1.1.2.1		Reextract. with 2 N H$_2$SO$_4$ in the aq. phase 2.2.1.2.1

Table 4.42. (continued)

	Aq. phase 2.2.1.2.1 extracted with E_1 at pH 3	
	Aq. phase 2.2.1.1.1	Org. phase 2.2.1.1.2
	A) Ile	Leu
	B) —	—
	C) —	—
	D) —	—
		Reextracted with H_2SO_4 into aq. phase

valine (Val). leucine (Leu), isoleucine (Ile), phenylalanine (Phe) and methionine (Met) in the first stage of Table 4.42), the high pK_3-value of the side group (at the separation of arginine (Arg) in the second stage), the different degrees of solvation (at the separation of Val and Met from Leu, Ile and Phe, as well as at the separation of Phe from Leu, and Ile, and finally at the separation of Leu and Ile), and the low pK_1-value (at the separation of histidine (His) from lysine (Lys)).

The main problem with these separation processes is our poor knowledge about the influence of the solvation, e.g., by the choice of the salt type (position in the Hoffmeister series: preservation or destruction of the water structure) and of the variation of the salt concentration. Also the interrelation between the complexes with the organic solvent and the modifier have not been investigated yet.

We need physical methods for the analysis of the composition of the associated complexes, their shape and size of aggregation as well as the influence of different salts on them. These are the prerequisites for the quantitative evaluation of the solvation effects [417].

References

References to Chapter 2

1. Sorensen JM, Artl W (1979–80) Liquid-liquid equilibrium data collection. Chemistry data series, vol V, pt 1–3, DECHEMA, Frankfurt
2. Flick EW (ed) (1985) Industrial solvents handbook. 3rd edn. Noyes Data Co, Park Ridge, N.J.
3. Kertes AS (ed) (1984) Alcohols with water. in: Solubility Data Series, vol 15, Barton AFM (ed) IPAC-Pergamon, Oxford
4. Wisniak J, Tamir A (1980) Liquid-liquid Equilibrium and Extraction. A literature Source Book. Physical Science Data No. 7, Parts A and B, Elsevier, Amsterdam
5. Kertes AS, King CJ (1986 March) Extraction chemistry of low molecular weight aliphatic alcohols. Technical Report LBL-21210, Lawrence berkeley Laboratory, University of California
6. Wardell JM, King CJ (1978) Solvent equilibria for extraction of carboxylic acids from water. Chem Eng Data 23: 144–148
7. Ricker NL, Michaels JN, King CJ (1979) Solvent properties of organic bases for extraction of acetic acid from water. J Separ Proc Technol 1: 36–41
8. Ricker NL, Pittmann EF, King JC (1980) Solvent extraction with amines for recovery of acetic acid. J Separ Technol 1: 23–30
9. Kertes AS, King CJ (1986) Extraction chemistry of fermentation product carboxylic acids. Biotechnol Bioeng 28: 269–282
10. Treybal RE (1963) Liquid Extraction. Mc Graw Hill Book Co, Inc, New York, 2nd Edition
11. Hanson C (ed) (1971) Recent Advances in Liquid-Liquid Extraction. Pergamon Press, Oxford
12. Ricci I (1951) The Phase Rule and Heterogeneous Equilibrium. D. van Nostrand Co, Inc., Princeton 1951
13. Bird RB, Steward WE, Lightfoot EN (1960) Transport Rhenomena. John Wiley & Sons, Inc, New York
14. Whitman WG (1953) Chem & Met Eng 29: 147
15. Levenspiel O (1962) Chemical Reaction Engineering, John Wiley and Sons, Inc, New York, London
16. Skelland AHP (1974) Diffusional Mass Transfer. John Wiley & Sons, New York
17a. Whitman WG (1923) Chem & Met Eng 29: 147
17b. Lewis WK, Whitman WG (1924) Ind Eng Chem 16: 1215
18. Higbie R (1935) Trans Am Inst Chem Engrs 31: 365
19. Danckwerts PV (1951) Ind Eng Chem 43: 1460
20. Danckwerts PV (1955) AIChE Journal 1: 456
21. Toor HL, Marcello JM (1958) AIChE Journal 4: 97
22. Carslaw HS, Jaeger JC (1947) Conduction of Heat in Solids. Oxford University Press, London
23. Newman AB (1931) Trans Am Inst Chem Engrs 27: 310
24. Kronig R, Brink JC (1950) Appl Sci Res A2: 142
25. Handlos AE, Baron T (1957) AIChE J 3: 127
26. Levich VG (1962) Physicochemical hydrodynamics. Prentice Hall, Engelwood Cliffs, N. J.
27. Hadamard JCR (1911) Comp Rend 152: 1735

28. Rybczynski W (1911) Bull Intern Acad Sci Cracovie (A) 40
29. Boussinesq J (1913) J Comp Rend 156 983, 1035
30. Savic P (1953) Mech Eng Rep MT-22 Natl Res Council of Canada
31. Brunson RJ, Wellek RM (1970) Can J Chem Eng 48: 267
32. Skelland AHP, Wellek RM (1964) AIChE Journal 10: 491 and 789
33. Rose PM, Kintner RC (1966) AIChE Journal 12: 530
34. Angelo JB, Lightfoot EN, Howard DW (1966) AIChE Journal 12: 751
35. Olander DR (1966) AIChE Journal 12: 1018
36. Hubis M, Hartland S (1986) Chem Eng Sci 41: 2436
37. Skelland AHP, Vasti NC (1985) Can J Chem Eng 63: 390
38a. Cruz-Pinto JJC (1979) PhD Thesis, Univ. Manchester, Inst Sci Technol, UK
38b. Korchinsky WJ, Cruz-Pinto JJC (1979) Chem Eng Sci 34: 551
39. Wellek RM, Skelland AHP (1965) AIChE Journal 11: 557
40. Patel JM, Wellek RM (1967) AIChE Journal 13: 384
41. Korchinsky WJ, Young CH (1986) Chem Eng Sci 41: 3053
42. Grace JR, Wairegi T, Nguyen H (1976) Trans Inst Chem Eng 54: 167
43. Garner FH, Tayeban M (1960) An Real Soc Esp Fis Quim Sec B Quim LVI (B) 479
44. Danckwerts PV (1970) Gas-Liquid Reactions. McGraw-Hill Book Co
45. Crank J (1956) The Mathematics of Diffusion. Clarendon Press, Oxford
46. Wellek RM, Andoe WV, Brunson RJ (1970) Can J Chem Eng 48: 645
47. Heertjes PM, Holve WA, Tasma H (1954) Chem Eng Sci 3: 122
48. Tyroler G, Hamielec AE, Johnson AI, Leclair BP (1971) Can J Chem Eng 49: 56
49. Halwachs W (1981) Habilitationsschrift, Universität, Hannover
50. Brunson RJ, Wellek RM (1971) AIChE Journal 17: 1123
51. Astarita G (1967) Mass Transfer with Chemical Reaction. Elsevier Publ Co, Amsterdam
52. Huang CJ, Kuo CH (1963) AIChE Journal 9: 161
53. Rod V (1974) Chem Eng J 7: 137
54. Reschke M, Halwachs W, Schügerl K (1982) Chem Eng Sci 37: 1529
55. Völkel W, Halwachs W, Schügerl K (1980) J Membrane Sci 6: 19
56. Griffith RM (1962) Chem Eng Sci 17: 1057
57. Davis RE, Acrivos A (1966) Chem Eng Sci 21: 681
58. Saville DA (1973) The Chem Eng Journal 5: 251
59. Linton M, Sutherland KL (1960) Chem Eng Sci 12: 214
60. Thorsen G (1954) PhD Thesis, Trondhein
61. Heertjes PH, De Nie LH (1971) in: "Recent Advances in Liquid-Liquid Extraction". Hanson C (ed) Pergamon Press, Oxford 367
62. Boussinesqu J (1905) J Math Pures Appl 60 (6): 285
63. Keey RTB, Glen JB (1965) Trans Inst Chem Engrs 43: 221
64. Weber ME (1975) Ind Eng Chem Fundam 14: 365
65. Harper JF, Moore DW (1968) J Fluid Mech 32: 367
66. Brauer H (1978) Int J Heat and Mess Transfer. 21: 445 and 455
67. Brounshtein BI, Fishbein GA (1974) Theor Found Chem Eng USSR 8(2): 186
68. Brauer H, Mewes D (1971) Stoffaustausch einschließlich chemischer Reaktion Verlag Sauerländer Aarau/Frankfurt
69. Skelland AHP (1974) Diffusional Mass Transfer. John Wiley & Sons, New York
70. Ruckenstein E (1967) Int J Heat and Mass Transfer 10: 1785
71. Gröber H (1925) Z Ver Dtsch Ing 69: 705
72. Gröber H, Erk S, Grigull U (1963) Grundgesetze der Wärmeübertragung, Springer Verlag, Berlin
73. Wellek R, Skelland AH (1965) AIChE Journal 11: 557
74. Schügerl K, Blaschke HG, Brunke U, Streicher R (1977) Interaction of Fluiddynamics. Interfacial Phenomena and Mass Transfer in Extraction Processes in "Recent Development in Separation Science". (ed) Li NN, vol. III. Part A, 71–129, CRC Press, Boca Raton Florida
75. Elzinga ER jr, Banchero JT (1959) Chem Eng Symp Progr Symp Ser 29: 149
76. Banerjee S, Lahey RT (1981) Adv Nucl Sci Technol 13
77. Chan AMC, Banerjee S (1981) Nucl Instrum Methods 190: 135
78. Buchholz R, Zakrzewski W, Schügerl K (1981) Internat Chem Eng 21: 180
79. Buchholz R, Tsepetonides J, Steinemann I, Onken U (1982) Chem Ing Techn 54: 840
80. Weiland P, Brentrup L, Onken U (1980) Ger Chem Eng 3: 296
81. Calderbank PH, Pereira J (1977) Chem Eng Sci 32: 1427

82. Popovich AT, Jervis RE, Trass O (1964) Chem Eng Sci 19: 357
83. Sawistowski H, Goltz GE (1963) Trans Inst Chem Engrs 41: 174
84. Coulson JM, Skinner SJ (1952) Chem Eng Sci 1: 197
85. Johnson AI, Hamielec AE (1960) AIChJ 6: 145
86. Garner FH, Skelland AHP (1954) Ind Eng Chem 46: 1255, (1955) Chem Eng Sci 4: 149, (1956) Ind Eng Chem 48: 51
87. Mensing W, Schügerl K (1970) Chem Ing Techn 12: 837
88. Halwachs W, Schügerl K (1982) Chem Eng Sci 38: 1073
89. Haensel R, Halwachs W, Schügerl K (1986) Chem Eng Sci 41: 555
90. Birks JB (1964) The Theory and Practice of Scintillation Counting. Pergamon, Oxford
91. Zimmermann V, Halwachs W, Schügerl K (1980) Chem Eng Commun 7: 95
92. Levenspiel O, Chem Reaction Engineering
93. Shah YT, Stiegel, GJ, Sharma HM (1978) AIChE Journal 24: 369–400
94. Levenspiel O, Bishoff KB (1963) Patterns of flow in chemical process vessels. in: "Advances in Chemical Engineering", Eds. TB Drew, JW Hoopes Jr, T Vermenlen, Acad, New York
95. Schügerl K, Verweilzeitverteilungen in Strömungssystemen. Fortschritt-Berichte, VDI-Zeitschrift, Reihe 7, Nr. 8

References to Chapter 3

96. LO TC, Bairds MHI, Hafez M, Hanson, C (eds) (1983) Handbook of solvent extraction. Wiley, New York
97. Hafez M, In Centrifugal Extractors. In Ref. 96, p. 459
98. Gebauer K, Steiner L, Hartland S (1982) Zentrifugalextraktoren–eine literaturübersicht. Chem Ing Techn 54: 476
99. Podbielniak WJ, Kaiser HR, Ziegenhorn GJ (1970) Centrifugal solvent extraction. In the history of pencillin production. Chem Eng Progr Symp Ser No. 100, 66: 45
100. Todd TB, Podbielniak WJ (1965) Advances in centrifugal extraction. Chem Eng Porgr 69
101. Anderson DW, Lau EF (1955) Commercial extraction of unfiltered fermentation broths in the Podbielniak contactor. Chem Eng Progr 51: 507
102. Godfrey JC, Slater MJ. Principles of mixer-settler design. In Ref. 96, p. 275
103. Lowes L. Simple box-type mixer settler. In Ref. 96, p. 279
104. Warwick GCI, Scuffham JB. The Davy-MC Kee mixer-settler. In Ref. 96, p. 287
105. Baenea E, Meyuer D. IMI mixer settlers. In Ref. 96, p. 299
106. Müller E, Stönner HM. Lurgi mixer-settlers. In Ref. 96, p. 311
107. Cavers SD. Nonmechanically agitated contactors. In Ref. 96, p. 319
108. Simons AJF. Pulsed packed columns. In Ref. 96, p. 343
109. Logsdail DH, Salter MJ. Pulsed perforated columns. In Ref. 96, p. 355
110. Lo TC, Prohaska J. Reciprocating-plate extraction columns. In Ref. 96, p. 373
111. Kostes WCG. Rotating-disc contactor. In Ref. 96, p. 391
112. Misek T, Marek J. Asymmetric rotating disc extractor. In Ref. 96, p. 407
113. Scheibel EG. Scheibel columns. In Ref. 96, p. 419
114. Oldshue JY. Oldshue-Rushton Column. In Ref. 96, p. 431
115. Mögli A, Buhlmann U. The Kühni extraction column. In Ref. 96, p. 441–447
116. Coleby J. The RTL (formerly Graesser raining-bucket) contactor. In Ref. 96, p. 449
117. Baird MHJ. Miscellaneous rotary-agitated extractors. In Ref. 96, p. 453
118. Pratt HRC, Hanson C. Selection, pilot testing, and scale up of commercial extractors. In Ref. 96, p. 475
119. Baird MHI, Lo TC. General laboratory scale and pilot plant extractors. In Ref. 96, p. 497
120. Perry JH, Perry RH, Chilton CH, Kirkpatrick SD (1963) Chemical Engineers' Handbook. McGraw-Hill Book Co, New York, 4th edn. Gas Absorption and Solvent Extraction. Emmert RE, Pigford RL. Section 14
121. Treybal RE (1972) Liquid-liquid Extraction. McGraw Hill Co, New York
122. Hanson C (1971) Recent Advances in Liquid-Liquid Extraction Pergamon, Oxford
123. Brandt HW, Reissinger K-H, Schröter J (1978) Chem Ing Techn 50: 345
124. Reissinger K-H, Schröter J, Bäcker W (1981) Chem Ing Techn 53: 607

125. Pilhofer T, Schröter J (1986) Ger Chem Eng 1: 1
126. Reissinger K-H, Marr R (1986) Chem Ing Techn 58: 540
127. Blass E, Goldmann G, Hirschmann K, Michailowitsch P, Pietsch W (1986) Ger Chem Eng 9: 222
128. Pilhofer T, Mewes D (1979) Siebboden-Extraktionskolonnen. Verlag Chemie, Weinheim
129. Brunner KH (1984) Separators and decanters for continuous extraction. Technical and scientific documentation No. 4, Westfalia Separator
130. The centrifugal Extractor ABE 216 and its use in the Pharmaceutical industry. Technical report. Alfa-Laval (p.p. 60926T)
131. Technical report, Robatel SLPI (Rue de Grenève/B.P. 203, F-69740 Genas
132. Technical Report, Liquid dynamics
133. Placek A (1933) US Patent 1 036 523
134. Podbielniak WJ (1935) US Patent 1 926 524
135. Placek A (1942) US Patent 2 281 616
136. Technical report, Baker Perkins. Inc., Scuite 217 2725N Thatcher Ave River Grove III 60171, USA
137. Doyle CM, Podbielniak-Doyle WG, Rauch EH, Lowry CD (1968) Chem Eng Progr 64 No. 12
138. Prohazka J, Landau J, Souhadra F (1970) US Patent 3 488 037
139. Landau J, Prohazka J, Souhadra F (1971) US Patent 3 583 856
140. Baird MHI, Mc Ginnis RG, Tan GC (1971) International Solvent Extraction Conference'71
141. Hafez MM, Baird MHI, Nirdosh I (1979) Can J of Chem Eng 57: 150
142. Baird MHI Lane SJ (1973) Chem Eng Sci 28: 947
143. Hafez MM, Baird MHI (1978) Trans Inst Chem Eng (London) 56: 229
144. Kim SD, Baird MHI (1976) Can J of Chem Eng 54: 81
145. Chem-Pro Equipment Bulletin KC-11, Karr Column, Fairland, N.J
146. Karr AE (1980) Separation Sci Technology 15: 877
147. Hafez MM, Baird MHI, Nirdosh I (1980) International Solvent Extraction Conference'80, Liege 80–41
148. Geier RG (1958) Proceedings of the 2nd conference on Peaceful Uses of Atomic Energy, Geneva 17, No. 515

References to Chapter 4

149. Falbe J, Payer W (1974) Alkohole. In: Ullmann's Encyklopädie der Technischen Chemie, Verlag Chemie, Weinheim, 7: 203
150. Kosaric N, Wieczorek A, Cosentino GP, Magee RJ, Prenosil JE (1983) Ethanol Fermentation. In: Rehm HJ, Reed G, Dellweg H (eds) Biotechnology, Vol 3, Verlag Chemie, Weinheim, 257
151. Maiorella BL (1985) Ethanol. In: Moo-Young M, Blanch HW, Drew S, Wang DIC (eds) Comprehensive biotechnology, Pergamon, Oxford 861
152a. Venkatasubramarian K, Keim CR (1981) Gasolhol: A commercial perspective. In: Vieth WR, Venkatasubramarian K, Constantinides A (eds) Biochem Engng II. New York Academy of Sciences 369: 187
152b. Keim CR (1983) Enzyme Micro Technol 5: 103
152c. Bremen L, Schmoltzi M (1986) Economics and politics of the ethanol market. TIBTECH, Jam. 16
153a. Walsh PK, Liu CP, Findley ME, Liapis AI, Siehr DJ (1983) Ethanol separation from water in a two-stage adsorption process. Biotechn Bioeng Symp No. 13: 629–647
153b. Pitt Jr WW, Haag GL, Lee DD (1983) Biotechn Bioeng 25: 123
153c. Malik RK, Ghosh P, Ghose TK (1983) Biotechn Bioeng 25: 2277
154a. Busche RM (1983) Recovering chemical products from dilute fermentation broths. Biotechn Bioeng Symp No. 13: 597–615
154b. Groot WJ, van den Oever CE, Kossen NWF (1984) Biotechnology Letters 6: 709.
154c. Asaeda M, Du LD, Fuji M (1986) J of Chem Eng of Japan 19: 84
155. Zacchi CL, Aly G, Wennersten R (1983) Investigation of ethanol extraction from fermentation liquids. ISEC'83, 507

156a. Munson CL, King CJ (1983) Factors influencing solvent selection for extraction of ethanol from aqueous solution. ISEC'83, 509
156b. Munson CL, King CL (1984) Ind Eng Chem Proc Des Dev 23: 109
157.   Ramalingham A, Finn RK (1977) Biotechnol Bioeng 19: 583
158.   Gysewsi GR, Wilke CR (1977) Biotechnol Bioeng 19: 1125
159.   Groot WJ, Schoutens GH, Van Beelen PN, Van den Oever CE, Kossen NWF (1984) Biotechnol Letters 6: 789
160.   Kollerup F, Daugulis AJ (1985) The Can J Chem Eng 63: 919
161.   Kollerup F, Daugulis AJ (1985) Biotechn Bioeng 27: 1335
162.   Kollerup F, Daugulis AJ (1986) The Can J of Chem Eng 64: 598
163.   Minier M, Goma G (1981) Biotechn Letters 3: 405
164.   Minier M, Goma G (1982) Biotechn Bioengng 29: 1565
165.   Gyamarach M, Glover J (1983) Ethanol by continuous fermentation using a combination of immobilized yeast and solvent extraction. Adv in Fermentation '83 Chelsea College, London (21–23 Sept 1983)
166.   Roddy JW (1981) Ind Eng Chem Proc Des Dev 20: 104
167.   Pye EK, Humphrey AE (1987, 1989) The biological production of liquid fuels from biomass. Univ Penn Annual Report to US Dept of Energy (Sept 1987–Sept 1979)
168.   Wang HY, Robinson FM, Le SL (1981) Enhanced alcohol production through on-line extraction. Biotechn Bioeng Symp (Third Biotechnol Energy Prod Conserv) 11: 555
169.   Finn RK (1966) J Ferm Technol 44: 305
170.   Levy S. Solvent extraction of alcohol from water solutions with fluorocarbon sovents. US Patent 4, 260, 836 (April 1981)
171.   Mattiasson, B, Hahn-Hägerdahl B, Albertson P (1981) Biotechn Letters 3: 53
172.   Eckles AJ, Ferster PJ, Tawfik WY, Tedder DW, Myerson AS (1984) Continuous fermentation and product recovery by liquid/liquid extraction (Poster Sea Island 1984)
173.   Tedder DW (1983) US Patent 4,399,000 (August 16, 1983)
174.   Matsumura M, Märkl H (1984) Appl Microbiol Biotechnol 20: 371
175.   Victor JG (1983) Extraction of ethanol from water with liquid propylene. ISEC'83, 511
176.   Paulaitis EM, Gilbert ML, Nahs CA (1981) Separation of ethanol-water mixtures with supercritical fluids. 2nd World Congress of Chem Eng, Montreal
177.   Kuk MS, Montagna JC (1983) Solubility of oxygenated hydrocarbons in supercritical carbon dioxide; Chemical engineering at supercritical conditions. Ann Arbor Sci 1983
178.   Brunner G, Kreim K (1985) Chem Ing Techn 57: 550 (MS 1371/85)
179.   Kreim K (1983) Dissertation, TU Hamburg-Harburg
180.   Brunner G (1987) Chem Ing Techn 59: 12
181.   Walton MT, Martin JL (1979) Production of Butanol-Acetone by fermentation in Microbial Technology, Vol I. HJ Pepper, D Perman (eds), 2nd Ed. Acad Press, London
182.   Eckert G (1986) Dissertation, Universität Hannover
183.   Dupire S, Thyrion FC (1986) Solvent extraction of products from Aceton Butanol fermentation. ISEC'86. Vol III, 613
184.   Essien DE, Pyle DL. Fermentation ethanol recovery by solvent extraction. Separations for Biotechnology. Verrall MS, Hudson MJ (eds) Ellis Horwood Ltd, Chichester, 32–332
185.   Tanaka H, Harada S, Kurosawa H, Yajima M (1987) Biotechnol Bioeng 30: 22–30
186   Atkinson B, Mavituna F (1983) Biochemical engineering and biotechnology handbook. Nature, McMillan, London
187.   Wagner FS Jr (1978) Acetic acid. In: Kirk-Othmer encyclopedia of chemical technology. 3rd edn., vol 1. Wiley-Interscience, New York, p 124
188.   Brockhaus R, Förster F (1976) Essigsäure. In: Ullmanns encyklopädie der technischen chemie 4th edn. vol 11. Verlag Chemie, Weinheim, p 57
189.   Eaglesfield P, Kelly BK, Short JF (1953) Ind Eng Chem 29: 243
190.   Brown WV (1963) Chem Eng Progr 59: 65
191.   Treybal RE (1963) Liquid Extraction. Section 15. In: Perry RH, Chilton CH (eds), Perry's Chemical Engineering Handbook. S.D. 5th edn, 1973, McGraw Hill, New York 15–18
192.   King CJ (1981) Acetic acid extraction. In: Lo TC, Baird MH, Hanson C (eds) Hand book of Solvent Extraction. Wiley – Interscience, New York, p 567
193.   Siebenhofer M, Marr R. Acid Extraction by Amines. ISEC'83, 219
194.   Siebenhofer M, Marr R (1985) Chem Ing Techn 57: 558
195.   Wojtech B, Mayer M (1985) Chem Ing Techn 57: 134

196. Shah DJ, Tiwari KK (1981) J Chem Eng Data 26: 375
197. Inoue K, Nakashio F (1974) Mass transfer accompanied by chemical reaction at the surface of a single droplet. Chem Eng Sci 34: 191
198. Pilhofer Th, Dichtl G. Liquid-liquid extraction as an alternative to distillation for the recovery of organic material. ISEC'86, III-723
199. Rückgewinnung von Essigsäure (1982) Quickfit-Corning Technical Bulletin and Chemie-Technik 11: 12
200. Grinstead RR (1974) US Patent 3 816 524 June 11
201. Baniel AM (1982) European Patent EP 49429
202. Baniel AM, Blumberg R, Hajdu K (1981) US Patent 4 275 234
203. Alter JE, Blumberg R (1981) US Patent 4 251 671
204. Wennersten R (1980) A new method for the purification of citric acid by liquid-liquid extraction. ISEC'80, Vol 2: 80–63
205. Wennersten R (1983) J Chem Techn Biotechnol 33B, 85
206. Jian Yu-Ming, Li Dao-Chen, Shu Yuan-Fu (1983) Study on extraction of citric acid, ISEC'83, 517
207. Friesen DT, Babcock WC, Chambers AR (1987) Separation of citric acid from fermentation beer using supported liquid membranes, Bend Research, Inc
208. Marvel CS, Richards JC (1949) Anal Chem 21: 1480
209. Smith EL, Page JE (1948) J Soc Chem Ing (London) 67: 48
210. Ratchford WP, Harris EH, Fisher CH, Willis CD (1951) Ind Eng Chem 43: 778
211. Pyatnitskii IV, Tabenskaya TV, Makarchuk TL (1973) J Analy Chem USSR 28: 484
212. Manenok GS, Korobanova VI, Yudina TN, Soldatov V (1979) Russ J Appl Chem 52: 156
213. Vieux AS, Rutagengwa N, Rulinda JB, Balikungeri A (1977) Anal Chim Acta 91: 359
214. Vieux AS, Rutagengwa N (1977) Anal Chim Acta 91: 359
215. Kawano Y, Kusano K, Inoue K, Nakashio F (1983) Kagaku Kogaku Ronbushu 9: 473
216. Kawano Y, Kusano K, Takashashi T, Kondo K, Nakashio F (1982) Kagaku Kogaku Ronbushu 8: 404
217. Kawano Y, Kusano K, Nakshio F (1983) Kagaku Kogaku Ronbushu 9: 211
218. Pyatnitskii IV, Tabenskaya TV (1970) J Anal Chem USSR 25: 2060
219. Pyatnitskii IV, Tabnskaya TV, Makarshuk TL (1973) J Anal Chem, USSR 28: 484
220. Pyatnitskii IV, Harchenko RS (1964) Ukr Khim Zhur 30: 635
221. Pyatnitskii IV, Harchenko RS (1963) Ukr Khim Zhur 29: 967
222. Pyatnitskii IV, Tabenskaya TV (1970) J Anal Chem USSR 25: 815
223. Pagel HA, McLafferty FW (1948) J Anal Chem 20: 272
224. Pagel HA, Schwab KD (1950) Anal Chem 22: 1207
225. Tamada JA, Kertes AS, King CJ (1986) Solvent extraction of succinic acid from aqueous solutions. ISEC'86, III-631
226. Rückl W, Siebenhofer M, Marr R (1986) Separation of citric acid from aqueous fermentation solutions by extraction reextraction process. ISEC'86, III-653
227. Voß, H (1985) Chem Ing Techn 57: 702
228. Czytko M, Ishii K, Kawai K (1987) Chem Ing Techn 59: 952
229   Nomura Y, Iwahra M, Hongo M (1987) Biotechnol Bioeng 30: 788
230. Lotz M, Czytko M (1990) Chem Eng Techn 62: 214
231. Rehmann D, Heyde M, Holley W, Bauer W (1992) Chem Ing Techn 64: 286
232. Bauer B, Chmiel H, Krumbholz C, Menzel Th, Schmidt K (1992) BIOforum 6: 202
233. Hano T, Matsumoto M, Ohtake T, Sasaki K, Hori F, Kawano Y (1992) Application of liquid membrane technique to the recovery of fermented organic acids. Sekine T (ed) ISEC'90, Kyoto, in Solvent Extraction, Part B, Elsevier, Amsterdam 1887
234. Bizek V, Horacek J, Kousova M, Heyberger A, Procházka J (1992) Chem Eng Sci 47: 1433
235. Boey S, Garcia del Cerro MC, Pyle DL (1987) Chem Eng Res Dev 65: 218
236. Siramm T, Pyle DL, Grandison AS (1990) In: "Separation for Biotechnology" Pyle DL (ed), SCI Elsevier, Amsterdam 245
237. Basu R, Sirkar KK (1992) Solvent Extraction and Ione Exchange, 10: 119
238. Frieling VP (1991) Doctoral Thesis, University Hannover
239. Holten CH (1971) Lactic acid, Verlag Chemie, Weinheim
240. Siebold M (1991) Doctoral Thesis, University Hannover
241. King CJ, Starr JN, Poole LJ, Tamada JA (1992) ISEC 90, Sekine T (ed) Solvent extraction 1990, Elsevier Sci Publ, Amsterdam, 1791

242. Yabannavar VM, Wang DIC (1991) Biotechn Bioeng 37: 1095
243. Joppien R (1992) Doctoral Thesis, University Hannover
244. Halwachs W (1981) Habilitationsschrift
245. Halwachs W, Schügerl K (1983) Chem Eng Sci 38: 1073
246. Schlichting E, Halwachs W, Schügerl K (1987) Chem Eng Commun 51: 193
247. Schlichting E, Halwachs W, Schügerl K (1985) Chem Eng Process 19: 317
248. Haensel R, Halwachs W, Schügerl K (1986) Chem Eng Sci 41: 135–141
249. Haensel R, Halwachs W, Schügerl K (1986) Chem Eng Sci 41: 555–565
250. Grosjean PRL, Sawitowski H (1980) Trans Inst Chem Engrs 58: 59
251. Otto W, Streicher R, Schügerl K (1973) Chem Eng Sci 28: 1777–1788
252. Streicher R, Schügerl K (1977) Chem Eng Sci 32: 23–33
253. Danckwerts PV (1970) Gas-Liquid Reactions. McGraw Hill Book Co New York
254. Kitai A, Ozaki A (1969) I Ferment Technol 47 527–535
255. Schmidt-Kastner G, Egerer P (1984) Amino acids and peptides. In: Rehm H-J, Reed G (eds)
     Biotechnology. A comprehensive treatise in 8 volumes. vol 6a, Biotransformations Kieslich
     K (vol ed) Verlag Chemie, Weinheim, p. 387
256. Yagodin GA, Yurtov EV, Golubkov AS (1986) ISEC'86, III, pp 677–683
257. Thien MP, Hatton TA, Wang DIC (1986) ISEC'86, III, pp 685–693
258. Behr J-P, Lehn JM (1973) J Am Chem Soc 95: 6108
259. von Frieling P (1988) Diploma thesis, University of Hannover
260. Haensel R, Halwachs W, Schügerl K, (1986) Chem Eng Sci 41: 1811–1815
261. Haensel R (1984) PhD thesis
262. Schlichting E, Halwachs W, Schügerl K (1987) Chem Eng Comm 51: 193–205
263. Schlichting E, Halwachs W, Schügerl K (1985) Chem Eng Process 19: 317–328
264. Kirgios I, Haensel R, Rhein HB, Schügerl K (1986) ISEC'86, III-623–629
265. Kirgios I, Rhein HB, Haensel R, Schügerl K (1986) Chem Ing Techn 58: 908–909
266. Kirgios I (1985) Diploma thesis, University of Hannover
267. Aliwarga L, Schügerl K (1987) Dechema Biotechnology Conferences 1: 483–488
268. Handojo L, Degener W, Schügerl K (1990) ISEC'90, Solvent Extraction. Sekine T (ed),
     Elsevier Sci Publ, Amsterdam 1785
269. Thien MP, Hatton TA, Wang DIC (1988) Biotechnol Bioeng 32: 604
270. Hatton TA (1987) Extraction of proteins and amino acids using reversed micelles. In: The use
     of Ordered Media in Chemical Separations, Hinze WL, Armstrong DW (eds) ACS Symp
     Series 342: 170
271. Scheper T, Halwachs W, Schügerl K (1983) ISEC'83
272. Scheper T, Halwachs W, Schügerl K (1984) The Chem Engng Journal 29: B31–B37
273. Hano T, Ohtake T, Matsumoto M, Kitamaya D, Hori F, Naksio F (1991) J Chem Eng of
     Japan 24: 20
274. Nozaki Y, Tanford C (1971) J Biol Chemistry 246: 221
275. Teramoto M, Miyake Y, Matsuyama H, Nara H (1990) ISEC'90 Solvent Extraction, Sekine
     T (ed) Elsevier Sci Publ, Amsterdam 1803
276. Hano T, Othake T, Matsumoto M, Kitamaya D, Hori F, Nakashio F (1990) ISEC'90,
     Solvent Extraction. Sekine T (ed) Elsevier Sci Publ, Amsterdam, 1881
277. Itoh H, Thien MP, Hatton TA, Wang DIC (1990) Biotechnol Bioeng 35: 853
278. Leonidis EB, Hatton TA (1990) J of Phys Chem 94: 6400
279. Leonidis EB, Hatton TA (1990) J of Phys Chem 94: 6411
280. Furusaki S, Kishi K (1990) J of Chem Eng Japan, 23: 91
281. Luisi PL, Laane C (1986) TIBTECH, June 153
282. Adachi M, Harada M, Shioi A, Sato Y, Uptake of amino acids into W/O-microemulsions J of
     Phys Chemistry (in press)
283. Atkinson B, Mavituna F (1983) Biochemical engineering and biotechnology handbook M,
     Nature, Macmillan, p. 984
284. Vandamme EJ (ed) (1984) Biotechnology of industrial antibiotics Marcel Dekker, New York
285. Vezina C, Sehgal SN (1984) Actinomycin. In: Ref. 284, p 629
286. White RJ, Stroshane RM (1984) Daunorubicin and adriamycin. In: Ref. 284, p 569
287. Miescher GM (1974) US Pat. 3 795 663
288. Vining LC, Westlake DWS , Chloramphenicol. In: Ref. 284, p 387
289. Butterworth D, Clavulanic acid. In: Ref. 284, p 225
290. Jost JL, Kominek LA, Hyatt GS, Wang HY, Cycloheximid. In: Ref. 284, p 531
291. Shreve RN, Brink JA (1977) Chemical Process Industries, McGraw Hill, New York, p 777

292. van Daehne W, Jahnsen S, Kirk I, Larsen R, Lörk H, Fusidic Acid. In: Ref. 284, p 427
293. Huber FM, Tietz AJ, Griseofulvin. In: Ref. 284, p 551
294. Higashide, Macrolides. In: Ref. 284, p 452
295. Rayman K, Hurst A, Nisin. In: Ref. 284, p 607
296. Sobin BA, Finlay AC, Kane JH (1977) US Pat 306 669
297. Hersbach GJM, Van der Beek CP, Van Dijk PWM, The penicillins. In: Ref. 284, p 45
298. Otake N, Salinomycin. In: Ref. 284, p 721
299. Podojil M, Blumauerova M, Vanek Z, Culik K, The tetracyclines. In: Ref. 284, p 259
300. Gray PP, Bhuwapathanapun S, Tylosin. In: Ref. 284, p 743
301. Biot AM, Virginiamycin. In: Ref. 284, p 695
302. Rigway K, Thorpe EE, Use of solvent extraction in pharmaceutical manufacturing processes,
      in Handbook of Solvent Extraction, Lo TC, Baird MHI, Hanson C (eds) Wiley New York
      1983, p 583
303. Kostareva MG, Eremina KK, Fedorova AN, Bityukskaya GA, Vyatkina MI, Vasil'era TL
      (1971) Antibiotiki (Moscow) 16(2): 134
304. Kostareva MG, Bedniagina NP, Bityuskaya GA (1974) Khim Pharm Zh 8(8): 44
305. Kostareva MG, Vyatkina NI (1977) USSR Pat 306 669 (Oct 25, 1977)
306. Rusin VN, Zhukovskaya SA, Pelbak VL (1975) Antibiotiki (Moscow) 20(3): 213
307. Swartz RW (1979) The use of economic analysis of penicillin G manufacturing costs in
      establishing priorities for fermentation process improvement. in: Ann Rep on Fermentation
      Processes. Vol 3, Acad Press, pp. 75–110
308. Sounders M, Pierotti GJ, Dunn CL (1970) The recovery of penicillin by extraction with a pH
      gradient. In: The history of penicillin production. (Chem Eng Prog Symp ser 100, vol 66)
      A.I.Ch.E. New York, p 37–42
309. Podbielniak WJ, Kaiser HR, Ziegenhorn GJ (1970) Centrifugal solvent extraction. In: The
      history of penicillin production.(Chem Eng Prog Symp Ser 100 vol 66) A.I.Ch.E. New York,
      p 45–50
310. Queener S, Swartz R (1979) Penicillins: Biosynthetic and semisynthetic. In: Economic
      microbiology, Vol 3, Rose AH (ed) Academic Press, New York, p 35–122
311. Rigdway K, Thorpe EE (1983) Use of solvent extraction in pharmaceutical manufacturing
      processes, in Handbook of Solvent Extraction Lo TC, Baird MHI, Hanson C (eds), New
      York 1983, p 583
312. Rowley D, Steiner H, Zimkin E (1946) Soc Chem Ind 65: 237
313. Kansava T, Kawai H, Ito T (1950) Takeda Res Lab 9: 57
314. Eisenlohr H, Scharlau A (1955) Pharmaz A, Ind 17: 207
315. Zhukowskaya SA (1972) Khim-Farm Zh 6(6): 42
316. Radzinskaya ESh, Rusin VN, Skobin MF, Zukovskaya SA, Izmailov RKh (1977) Khim
      Farm Zh 11(12): 77
317. Katinger H, Wibbelt F, Scherfler H (1981) vt-verfahrenstechnik 15: 179
318. Westfalia Separator AG, Oelde. Operational Manual No 8194-9000-001. Print 0585
319. Brunner KH, Scherfler H, Stölting M (1981) vt-verfahrenstechnik 14: 619
320. Reschke M, Schügerl K (1984) The Chem Eng Journal 28: B1–B9
321. Reschke M, Schügerl K (1984) The Chem Eng Journal 28: B11–B20
322. Reschke M, Schügerl K (1984) The Chem Eng Journal 29: B25–B29
323. Reschke M, Schügerl K (1985) The Chem Eng Journal 31: B19–B26
324. Reschke M, Schügerl K (1986) The Chem Eng Journal 32: B1–B5
325. Müller B, Schlichting E, Bischoff L, Schügerl K (1987) Appl Microbiol Biotechnol 26: 36–41
326. Müller B, Schlichting E, Bischoff L, Schügerl K (1987) Appl Microbiol Biotechnol 26:
      206–210
327. Müller B, Schlichting E, Schügerl K (1988) Appl Microbiol Biotechnol 27: 484–486
328. Likidis Z, Schügerl K (1987) Journal of Biotechnol 5: 293–303
329. Likidis Z, Schügerl K (1988) Bioprocess Engineering 3: 79–82
330. Likidis Z, Schügerl K (1988) Chem Eng Process 23: 61–64
331. Likidis Z, Schügerl K (1988) Chem Eng Sci 43: 27–32
332. Likidis Z, Schügerl K (1988) Chem Eng Sci 43: 1243–1246
333. Likidis Z, Schügerl K (1987) Biotechnol Letters 9: 229–232
334. Likidis Z, Schügerl K (1987) Biotechnol Bioengng 30: 1032–1040
335. Likidis Z, Schlichting E, Bischoff L, Schügerl K (1989) Biotechnol Bioengng 33: 1385–1392
336. Likidis Z, Reschke M, Scheper T (1987) Chem Ing Techn 59: 522–523
337. Reschke M (1983) PhD Thesis, University of Hannover

338   Müller, B., PhD Thesis, 1985, University of Hannover
339. Likidis Z (1986) PhD Thesis, University of Hannover
340. Brunner KH, private communication
341. Makryaleas K, Scheper T, Schügerl K, Kula M-R (1985) Ger Chem Eng 8: 345–350
342. Scheper T, Makryaleas K, Nowottny C, Likidis Z, Tsikas D, Schügerl K (1987) Enzyme Engineering 8, Annals of the New York. Acad Sci Vol 501: 165–170
343. Scheper T, Likidis Z, Makryaleas K, Nowottny C (1987) Schügerl K, Enzyme Microb Technol 9: 625–631
344. Margreiter H, Gapp F (1962) Penicillin-Chemie und Eigenschaften. In: Die Antibiotika. Brunner R, Machacek G (eds) Vol 1, Part I, Hans Carl, Nürnberg, p 225
345. Beecham Group Ltd British Pat 1 565 656 (Dec 13, 1975)
346. Wilke CR, Chang P (1955) A.I.Ch.E. Journal 1: 264–270
347. Chapters 9.1 to 9.5 (1983) In: Handbook of Solvent Extraction. Lo TC, Baird MHI, Hanson C (eds) New York
348. Hafez M (1983) Centrifugal Extractors. In: Handbook of Solvent Extraction Lo TC, Baird MHI, Hanson C (eds) Wiley, New York, p 459
349. Behr JP, Lehn JM (1973) J Am Chem Soc 95; 18 Sept 1973, 6108–6110
350. Li NN (1968) US Pat 3410794
351. Li NN, Cahn RP, Shrier AL (1971) US Pat 3617546
352. Li NN, Cahn RP (1973) US Pat 3719590
353. Halwachs W, Schügerl K (1978) Chem Ing Tech 50: 767–774
354. Barenschee T. Scheper T, Schügerl K (1992) J of Biotechnology 26: 143–154
355. Pitt WW Jr, Haag GL, Lee DD (1983) Biotechnol Bioeng 25: 123
356. Nielsen L, Larsson M, Holst O, Mattiasson B (1988) Appl Microbiol Biotechnol 28: 335
357. Dykstra KH, Li X-M, Wang HY (1988) Biotechnol Bioeng 32: 356
358. Roychoudry PK, Ghose TK, Ghosh P, Chotani GK (1986) Biotechnol Bioeng 28: 927
359. Podojil M, Blumauerova M, Vanek Z, Culik K (1984) The tetracyclines: properties, biosynthesis, and fermentation. In: Vandamme EJ (ed) Biotechnology of industrial antibiotics. Marcel Dekker, New York, Basel 259
360. Lee CK, Hong J (1988) Biotechnol Bioeng 32: 647
361. Joppien R (1992) Dissertation, University Hannover
362. Pons MN, Tillier-Dorion F, Gas membrane removal of volatile inhibitors, Proc. 4th European Congress on Biotechnol Niessel OM, van der Meer RR, Luyben KChAM (eds) Elsevier Science Publ B.V. Amsterdam Vol I, 149
363. Gudernatsch W, Mucha H, Hofmann Th, Strathmann H, Chmiel H (1989) Ber Bunsenges Phys Chem 93: 1032
364. Mulder MHV, Smolders CA (1986) Process Biochemistry, April 35
365. Groot WJ, Schoutens GH, Van Beelen PN, Van den Over PN, Kossen NWF (1984) Biotechnol Letters, 6: 789
366. Groot WJ, van der Laans RGJM, Luyben KChAM (1989) Appl Microbiol Biotechnol 32: 305
367. Lechner M, Märkl H, Götz F (1988) Appl Microbiol Biotechnol 28: 345
368. Choudhry JP, Ghosh P, Guha BK (1985) Biotechnol Bioeng 27: 1081
369. Daugulis AJ, Swaine DE, Kollerup F, Groom CA (1987) Biotechnology Letters 9: 425
370. Job C, Blass E, Schertler C, Staudenbauer WL (1989) Ber Bunsenges Phys Chem 93: 997
371. Bruce LJ, Daugulis AJ (1992) Biotechnol Letters 14: 71
372. Daugulis AJ, Axford DB, McLellan PJ (1991) The Can. J Chem Eng 69: 488
373. Bruce LJ, Axford DB, Ciszek B, Daugulis AJ (1991) Biotechnol. Letters 13: 291
374. Matsumura M, Märkl H (1984) Appl Microbiol Biotechnol, 20: 371
375. Gianetto A, Ruggeri B, Specchia V, Sassi G, Forna R (1988) Chem Eng Sci 43: 1891
376. Honda H, Taya M, Kobayashi T (1986) J of Chem Eng Japan 19: 268
377. Tanaka H, Harada S, Kurosawa H, Yajima M (1987) Biotechnol Bioeng 30: 22
378. Aires Barros MR, Cabral JMS, Novais JM (1987) Biotechnol Bioeng 29: 1097
379. Oliveira AC, Cabral JMS (1991) J Chem Tech Biotechnol 52: 219
380. Vatai Gy, Tekic M (1991) Sep Sci and Technol, 26: 1005
381. Bandini S, Gostoli C (1990) Ethanol production by extractive fermentation: A novel membrane extraction technique. In: Separation for Biotechnology Pyle DL (ed) SCI/Elsevier Amsterdam, 539
382. Christen P, Minier M, Renon H (1990) Biotechnol Bioeng, 36: 116
383. Fournier RL (1988) Biotechnol Bioeng 31: 235

384. Naser S, Fournier RL (1988) Biotechnol Bioeng 32: 628
385. Tedder DW, Tawfik WY, Poehlein SR (1986) Ethanol recovery from low-grade fermentates by solvent extraction and extractive distillation: The seed process ISEC'86, III-659
386. Udriot H, Ampuero S, Marisol IW, von Stockar U (1989) Biotechnol Letters 11: 509
387. Eckert G, Schügerl K (1987) Appl Microbiol Biotechnol, 27: 221
388. Wayman M, Parekh R (1987) J Ferment Technol 65: 295
389. Barton WE, Daugulis AJ (1992) Appl Microbiol Biotechnol, 36: 632
390. Roffler SR, Blanch HW, Wilke CR (1988) Biotechnol Bioeng, 31: 135
391. Jeon YJ, Lee YY (1989) Enzyme Microb Technol 11: 575
392. Roffler S, Blanch HW, Wilke CR (1987) Biotechnol Progress 3: 131
393. Groot WJ, Timmer JMK, Luyben KChAM (1987) Membrane solvent extraction for in-situ butanol recovery in fermentations, Proc 4th European Congress on Biotechnology, vol 3, Neissel OM, van der Meer RR, Luyben KChAM (eds) Elsevier Sci Publ B.V. Amsterdam, 564
394. Shukla R, Kang W, Sirkar KK (1989) Biotechnol Bioeng 43: 1158
395. Matsumura M, Takehara S, Kataoka H (1992) Biotechnol Bioeng 39: 148
396. Groot WJ, van der Lans RGJM, Luyben KChAM (1992) Process Biochemistry, 27: 61
397. Yabannavar VN, Wang DIC (1987) Bioreactor System with Solvent Extraction for Organic Acid Production, Annals New York Academy of Sciences, 523
398. Yabannavar VM, Wang DIC (1991) Biotechnol Bioeng 37: 544
399. Yabannavar VM, Wang DIC (1991) Biotechnol Bioeng 37: 716
400. Yabannavar VM, Wang DIC (1991) Biotechnol Bioeng 37: 1095
401. Bar R, Gainer JL (1987) Biotechnol Progress 3: 109
402. Bartels PV, Drost JCG, de Graauw J (1987) Solvent extraction of carboxylic acids from aqueous solution using porous membrane module, Proc 4th European Congress on Biotechnol Vol 2, Neissel OM, Van der Meer RR, Luyben KChAM (eds) Elsevier Science Publ B.V. Amsterdam, 558
403. Prasad R, Sirkar KK (1987) A.I.Ch.E. Journal, 33: 1057
404. Aires-Barros MR, Cabral JMS, Wilson RC, Hamel JFP, Cooney CL (1989) Biotechnol Bioeng. 34: 909
405. Siebold M, Frieling VP, Joppien R, Rindfleisch D, Schügerl K, Röper H (1994) Process Biochemistry (in press)
406. Frieling VP (1991) Entwicklung und Characterisierung von Systemen für die Reaktivextraction von Milchsäure. Doctoral Thesis, University Hannover
407. England R, Abed-Ali SS, Brisdon BJ (1988) Direct extraction of biotransformation products using functionalised polysiloxanes, Dechema Biotechnol Conf. Vol 2, VCH, Weinheim 71-81
408. Mavituna F, Wilkinson AK, Williams PD (1987) Liquid-liquid extraction of plant secondary metabolite as an integrated stage with bioreactor operation. Separations for Biotechnol. Verral MS, Hudson MJ, Ellis Horwood, 333-339
409. Hollmann D, Merrettig-Bruns U, Müller U, Onken U (1990) Secondary metabolites by extractive fermentation, Separation for Biotechnology, Pyle DL, SCI/Elsevier, 567-576
410. Müller U, Träger M, Onken U (1989) Ber Bunsenges Phys Chemie 93: 1001-1004
411. Barenschee T, Scheper T, Schügerl K (1992) J of Biotechnol 26: 143-154
412. Rindfleisch D (1994) University Hannover, Dissertation
413. Kirgios I, Rhein HB, Haensel R, Schügerl K (1986) Chem Ing Techn 58: 908
414. Handojo L (1988) Dissertation, University of Hannover
415. Frieling PV (1988) Diploma thesis, University of Hannover
416. Bitar MC, Sabot JL, Aviron-Violett P (1987) E-Pat P 0251 852, C 07 C 99/02 (12.06.1987) (Rhone-Poulenc Chimie)
417. Schügerl K, Degener W (1989) Chem Ing Techn 61: 796-804
418. Brandes A (1987) Diploma thesis, University of Hannover

Subject Index

M.S. Oka, **R.G. Rupp** (Eds.)

Cell Biology and Biotechnology

Novel Approaches to Increased Cellular Productivity

1993. Approx. 175 pp. 47 figs. (Serono Symposia) ISBN 3-540-97951-4

Cell Biology and Biotechnology: Novel Approaches to Increased Cellular Productivity contains the proceedings of the symposium by the same name held in Cambridge, Massachusetts, January 30 - February 2, 1992. State-of-the-art research is presented on: the IGF-1 Receptor and Gene Expression During the Cell Cycle; Attachment Control of Fibroplast Proliferation; Cell Biology and Serum-Free Mouse Embryo Cells; Immunoglobulin Production Stimulating Factors; Erythropoietin Control of Programmed Death in Erythroid Progenitors; Prohormone Processing Enzymes and Protein Production; Control of Translation Initiation by Phosphorylation; Protein Retention in the Endoplasmic Reticulum Mediated by GPR78; Molecular Approaches Towards Manipulating the Expression of the Glucose Related Proteins in Mammalian Cells; Protein Folding in the Endoplasmic Reticulum; Sorting of Membrane Proteins in the Endocytic and Exocytic Pathways; CIS-Acting Elements which Regulate Immunoglobulin Gene Transcription.

 Springer

Tm.BA94.004

K. Faber

Biotransformations in Organic Chemistry

1992. X, 319 pp. 31 figs. 238 schemes, 18 tabs. ISBN 3-540-55762-8

The book describes the use of biocatalysts (i.e. enzymes and/or whole cells) for organic synthesis. It is the first approach of a condensed introduction into this field which has mainly been developed during the past ten years. Emphasis is laid on chemo-, regio- and stereoselective transformations of non-natural organic compounds; products are e.g. pharma- ceuticals and fine chemicals. This professional referencebook is aimed at graduate students and scientists in organic chemistry and biotechnology as well as professionals in the fine chemicals, pharmaceutical and food(additives) industry.

R.K. Scopes

Protein Purification

Principles and Practice

3rd ed. 1993. Approx. 350 pp. 165 figs. (Springer Advanced Texts in Chemistry) ISBN 3-540-94072-3

The third edition of this classic guide to protein purification updates methods, principles and references.

As in the widely-acclaimed earlier editions, Scopes guides both the novice and the experienced researcher from theory to application. Using the book, the reader will be able to integrate methods effectively into optimum protocols for the task at hand.

Tm.BA94.004